2011

广州土人景观效果图表现十年集

GUANGZHOU TURENSCAPE RENDERING PROTFOLIO 10 YEARS

黄志坚 李津 等著

景观 LANDSCAPE 设计 DESIGN　　大连理工大学出版社

图书在版编目（CIP）数据

广州土人景观效果图表现十年集：2001~2011：汉英对照 / 黄志坚等著. -- 大连：大连理工大学出版社，2012.4

ISBN 978-7-5611-6848-6

Ⅰ.①广… Ⅱ.①黄… Ⅲ.①景观设计—作品集—中国—现代 Ⅳ.①TU986.2

中国版本图书馆CIP数据核字(2012)第062380号

出版发行：大连理工大学出版社
　　　　　（地址：大连市软件园路80号 邮编：116023）
印　　刷：利丰雅高印刷（深圳）有限公司
幅面尺寸：215mm×265mm
印　　张：14
字　　数：14千字
出版时间：2012年5月第1版
印刷时间：2012年5月第1次印刷
策划编辑：苗慧珠
责任编辑：刘晓晶
责任校对：周　阳
版式设计：王　江　黄志坚

ISBN 978-7-5611-6848-6
定　价：　198.00元

电　话：0411-84708842
传　真：0411-84701466
邮　购：0411-84708943
E-mail:dutp@dutp.cn
http://www.landscapedesign.net.cn

入设计行，天天琢磨着怎么画，怎么画好。其实，更重要的事应该是怎么想、怎么做；同等重要的事还包括怎么看、怎么说、怎么写、怎么处（同事、甲方等）。设计是个十分复合的工作。

有些人设计学不好，以为是美术不好，画得不好。不对，设计不好就是设计不好。设计好可以画得好，也可以画得不好。你能画清楚其实就可以。毕竟，设计不是画画，设计师也不是画家。

我们的创造力、原创力低下从画画就开始了。不鼓励真正的自主观察、自主表现，范画的力量过大，画出来都太像。画得跟老师一样，画得跟书上一样就是好画，有没有自己的观察变得毫不重要，谬矣。

效果图是设计的直观表达、通俗表达，是一套图纸里最能与非专业人士沟通的部分。你要重视和致力于沟通，你要赢得非专业人士对设计的认同、理解。至少，在现阶段我们还真要画好效果图。

效果图里就有设计思想，就有设计观念，就有设计师的立场和出发点。

是自己人画和还是发包请专业表现公司画，我们倾向前者。前者懂啊，自己人懂自己的活。自己的人画就可以把琢磨设计和画效果图结合起来做，弄在一起做。

当然，公司效果图画得越多，画得越好，我们就越知道效果图的局限和弊端。我们期待做更多的模型，期待用更踏实靠谱的方式表达和琢磨设计。

对设计效果图过分重视，说明我们的社会太过重视设计的视觉性。设计有许多非视觉性的好处，效果图是表达不出来的。过分重视效果图的效果，一定程度影响了社会（包括我们自身）更全面地认识设计。

效果图还得画下去，最近广土规定：在本国境内的项目，设计效果图上的人物不能画一堆外国人，要画当地人！画图的人，潜意识也好，明意识也好，你要明白你设计的这个东西给谁用！谁来用！这也算我们这些人的执着或老土吧。

庞伟
北京土人景观与建筑规划设计研究院 副院长
广州土人景观顾问有限公司 总经理兼首席设计师
北京大学景观设计研究院 客座研究员
广州美术学院设计学院 客座教授
2012 年 3 月 1 日

At the beginning of my designing career, what I focused on was drawing and trying to draw better. But later, I found that thinking and doing are more essential, and they are as important as how to look, how to talk, how to write, and how to deal with others like colleges and clients. Design is a complex job. Some people hold the opinion that student of poor design is due to his poor ability of arts and drawing. I cannot agree. Bad design is bad design. A good design has nothing to do with drawing well or bad as long as you draw it out clearly. All in all, designing is not drawing, and a designer is not a painter.

The inadequate creativity and originality begins from painting. Laying stress on model rather than independent observation and performance leads to the similar paintings. It will make a big mistake if we ignore the observation and believe that the more you draw like the teacher or the book, the better a painting is.

The rendering tells the design concept directly and simply. It is the only part that can communicate with non specialists in the whole blue print. To get the recognition and understanding of non specialists, we have to focus on communication, which means, we still have to draw a good rendering at least for the time being.

It contains the design idea, design concept, and the designer's standpoint and starting point in the rendering.

Painting it by ourselves or by professional company, we prefer the former. We know what we want, and we can combine drawing renderings with designing process.

Of course, the more and better renderings the company provides, the more limitations and inadequacies we will find. We anticipate more models and more practical ways of expressing and studying design.

It shows that our society pays too much attention on visuality of design that we overvalue the renderings. In fact, there are so many non-visual advantages of design that the renderings cannot perform. Overvaluing the effect of the print, at a certain point, prevents the society including ourselves recognizing the design in a more comprehensive view.

We still need to draw renderings. Guangzhou Turen Landscape CO, LTD published a new rule recently that there should be Location people rather than foreigners in the design renderings for domestic project. The people, who paint the rendering, have to know about whom you are design for and who will use what you design. You can consider us too stubborn or old-fashioned.

Wei Pang

Deputy President of Bejing Turen Landscape and Architecture Design Institute

General Manager & Chief Designer of Guangzhou Turen Landscape Planning CO, LTD

Guest Researcher of the Graduate School of Landscape Architecture, Peking University

Guest Professor of College of Design of Guangzhou Academy of Fine Arts

Mar 1st, 2012

为了设计而表现

Rendition for Design

　　广州土人团队自 2000 年组建至今已 12 年，在严酷的市场生态中历练并成长，健康甚至有点"野"。这本书能从局部反映公司的发展历程，虽然只是一本画册，但里头包含着大量的信息和故事。

　　我们这个团队的工作方法是在首席设计师庞伟老师的带领下，经过 12 年的积累总结形成的；所以很有效也很独特，我们对设计表现的理解和应用都体现了这一点。早在公司成立初期的岐江公园项目中，一张优秀的效果图曾为项目的进展起到过很大的推动作用。据老庞回忆：在紧迫的工期压力下，公园中那座著名的灯塔的设计方案迟迟无法定案，最终在一位很有素养的电脑效果图工作者的帮助下，绘制了一张很有说服力的灯塔效果图，使设计方案在评审会上一致通过。灯塔建成之后，达到了预想效果并得到了广泛认可。这一事件，可以算是广州土人设计表现工作的开端，在之后的工作中，逐渐形成了设计和表现一体化的工作方式，效果图全部由设计师们完成。

　　我们的效果图风格也一直在变化。2000~2003 年间的效果图比较抽象，着重表现空间结构，比如佛山调蓄湖、南海狮山文化中心、顺德新城区文化四馆等项目。2004~2006 年，着重表现景观意境和材质，比如顺德碧水商城、佛山梁园规划、广州光大花园等项目。2007~2009 年，着重表现氛围、植物，比如东莞黄旗山公园规划、美的总部大楼景观等项目。2010 年至今，多种风格并存，把更多的主张融入到效果图中，比如广州北岸码头和热电厂改造、四会农业园、大岭山湿地公园等项目。12 年间，虽然也不断有人员流动，但设计表现工作并未停止过前进，必须感谢本书的两位作者。黄志坚自 2002 年加入团队以来，一直是公司设计表现工作的核心人物，他开创了广州土人设计表现的风格。李津自 2005 年加入团队以来，开创了高效的效果图制作方法，并促进了表现风格的多样性。首席设计师和设计总监们对设计表现提出的要求都比较高，这也在一方面促使了设计师们进步；设计师们进步了，公司则会对设计表现提出更高的要求。

　　设计表现一直是存在争议的，许多业内人士认为效果图带有欺骗性，背离了设计的本质。设计表现被贴上了"取巧"和"忽悠"的标签，这使得设计师们在宣扬设计表现时不太理直气壮。实际上，这不是设计表现的问题，而是设计表现工作者的问题。设计表现的作用大致有两种：一种用于畅想和憧憬，画的是远期的场景或对可能性的探讨，多是不真实的；另一种是用于推敲和预演设计的，必须忠实于设计。因为设计工作追求的是优秀的建成作品，而不是优秀的效果图作品，所以设计师不能用效果图欺骗自己、欺骗建设方。从我们的实践经验看，把设计表现作为推敲和预演设计的工具是有效的。深圳中科研发园、东部华侨城湿地花园、美的总部大楼景观、共和生态公园等建成项目的效果，基本和方案阶段效果图的预演是一致的。当然，每个项目的推敲都不可能一步成型，都会由首席设计师牵头经过多次的修改，设计表现也都要同步跟进，有时反复的次数会很多。在这个过程中，设计师们对项目的理解会逐渐加深，对设计的把握也会得到有效的训练。

　　广州土人的效果图以电脑绘制为主，主要是因为便于用电子模型推敲设计。当然，也不排斥手绘效果图，因为对景观植物场景的表现，手绘有时更有优势。以黄志坚和李津为首，肩负着设计表现任务的设计师们的成长，体现出了公司的价值观。黄志坚在大学所学的专业是机械设计，他未曾接受过正规的计算机以及美术方面的训练，加入团队以来，凭借着对这个行业的热爱，不辱使命，制作了大量优秀的效果图和平面设计作品；黄志坚平时少言寡语，但做图总能让人眼前一亮，为公司的许多重要项目立下了汗马功劳。李津加入团队时几乎没有基础，同样是在自身的动力和公司所提要求的共同作用下迅速成长起来的；李津风风火火、高谈阔论、又精于钻研，他的工作效率在一定程度上改变了公司的工作方式，至今仍保持着公司单位时间内作图量最高的纪录。他俩不单是"土"人，还是"野"人——这两位在公司里早已成为了偶像级人物。

　　在此，还要对为广州土人设计表现工作做出贡献的设计师们表示感谢！

张健
广州土人景观顾问有限公司 董事、设计总监
景观设计师
《景观设计》杂志编委
2012 年 2 月 29 日

Over the last 12 years since its establishment in 2000, Guangzhou Turen Landscape Planning CO, LTD has tempered itself in the severe marketing environment, with a momentum that is "wild" even. This book, though a picture album at most, reflects the development process of the company in a way and contains a good amount of information and stories in it.

Formed from the accumulated experience of 12 years under the guidance of our chief designer Wei Pang, the working method of our team is quite unique and efficient, which is evidenced by our special understanding and application of the design rendering. In the Shipyard Park project in the early years of our company, a creative design rendering has played a huge propelling role in the progress of the project. According to Mr. Pang's retrospect, the design scheme of a landmark of a lighthouse in the park remained unfinished under the deadline pressure precariously when a satisfying lighthouse rendering was produced with the help of an adroit computer effects worker and this design was applauded and passed unanimously. When the lighthouse was finished, the former anticipation was gratifyingly met and we received widespread recognition. This successful project can be deemed as the beginning of our design rendering work; in subsequent work, we gradually attained the working method of unifying design and rendition and the rendering was completed by the designers themselves independently.

The style of our rendering keeps changing all the time. The rendering between 2000 and 2003 was inclined to be abstract, laying emphasis on the spatial structure, as was demonstrated in such projects as the Foshan Storage Lake, the Nanhai Shishan Cultural Centre and the Cultural Four Halls of the new city district of Shunde. In the span between 2000 and 2003, the landscape mood and materials quality was put into priority, as was demonstrated in such projects as the Shunde Bishui Mall, Foshan Liangyuan Garden and Guangzhou Guangda Garden, while between 2007 and 2009, the elements of atmosphere and plants were highly valued, as was demonstrated in such projects as Planning for Huangqi Mt. City Park, Dongguan and the headquarters of the company of Midea. From 2010 till now, diverse styles coexist and more elements have been integrated into the rendering, as was illustrated in such projects as the Guangzhou North Bank Wharf and Thermoelectric Plant, Sihui Agricultural Garden and Da Lingshan Wetland Park. In the past 12 years, despite the incessant personnel turnover, the design rendering work has never ceased to make progress. Special acknowledgements must be paid to the two authors of this book. One is Zhijian Huang who, since his joining our team in 2002, has always been the core figure in the design rendering work of our company and has acted as the founding father of the design rendering style of our company. The other one is Jin Li who, since his joining our team in 2005, has launched an efficient design rendering method and improved the diversity of rendering styles. On one hand, the chief designer and design director have posed a high benchmark against the design rendering, spurring the designers on. On the other hand, the designers' improvement would induce the company to raise the demand to an even higher level.

Design rendering has long been a controversy-provoking subject. Many insiders believe that the design rendering is misleading and goes against the nature of design. Such labels as "contrived artfulness" and "deceptive" put onto design rendering have greatly reduced the designers' convincingness when they are publicizing it. As a matter of fact, it is owing to the own problems of the design rendering workers rather than design rendering per se. Design rendering generally has two effects. One is to feed imagination. Design rendering for this end invariably deals with a distant scene or the ranges of possibility, tending to be surreal. The other one, which is intended for analyzing and rehearsing the actual designs, must follow the design on hot heel, as it's the excellent materialized work rather than the refined design rendering itself that the design work pursues. Therefore, the designer must not mislead himself and the construction party with the rendering. According to our practical experience, it's very effective to use design rendering as a tool for analyzing and rehearsing the actual designs. In such projects as Shenzhen Zhongke R&D Garden, East OCT Wetland Garden, the headquarters of the company of Midea and Gonghe Ecological Park, the actual results were by and large in accordance with the design rendering. As a matter of course, the analysis of the project design could not attain perfection on first trial and would need numerous corrections directed by the chief designer. Though in keeping in pace, the corrections may probably retrace their steps over and again, yet the designers will henceforth have their understanding of the project deepened and get effectively trained in grasping designs as a whole.

The rendering of Guangzhou Turen Landscape Planning CO, LTD is usually computerized, mainly considering the convenience in using electronic models to analyze the design. Doubtlessly, handmade rendering is also employed for its particular advantage in displaying the garden plants scene. Headed by Zhijian Huang and Jin Li, the designer team's growth in design rendering tasks mirrors the corporate values. Zhijian Huang majored in mechanical designing in college and has never received any official training in computers or drawing. Since his joining the team, inspired by his ferventness for this industry, he has made large numbers of excellent rendering and works of graphic design, contributing immensely to many of the important corporate projects. A reticent person himself, he, however, can always make you gape at his works. When Jin Li first joined the team, he was without any professional background but grew rapidly under the combined driving force of his own determination and the company's demand. He was swift, knowledgeably talkative and research-bent. To some degree, his efficient manner changed the way of work of the company. Up till now, he still holds the record of making the most rendering pictures in a given period in the company. They two are not only "ingenuous" as the company name translates but also "wild", of course in terms of their working manners. Both of them have already become icons in the company.

Here acknowledgements are extended to all the designers (including those who have left the office) who have made their share of contribution to Guangzhou Turen Landscape Planning CO, LTD.

Jian Zhang

Director and Design Director of Guangzhou Turen Landscape Planning CO, LTD

Landscape Designer

Member of Editorial Committee of Landscape Design Magazine

February 29th, 2012

在设计表现中思考

Reflection in the Process of Design Rendering

　　设计强调以"图"说话，用"图"表述设计理念，设计表现是贯穿设计整个过程不可缺失的重要工具。因此，设计效果图的表现技法在景观设计教育与实践中，成为设计师及准设计师们刻苦磨练、反复研习的"基本功"。设计效果图形式多样：或是概念形成的初期草案，或是思维过程中的推敲手稿，或是最终成果的设计文件。它最常见的形式是对三维场景模拟表现，由于直观、便于非专业人士理解设计，这种类型设计效果图在设计使用中最为普及，也在本书中占有主要篇幅。除此之外，设计表现方式还包括矢量化的CAD图示，内容有反映布局位置关系的平面图与总平面图、反映竖向关系的剖面图、反映设计结构关系的各类分析图、反映外型效果的立面图等，这些抽象的表现形式主要用在专业技术的设计图纸上。

　　在中国，设计效果图制作已发展成为一个专业化、市场化的行业，效果图制作师独立于设计师，可以是非设计专业人士，可以是流水线的工作方式。通常的工作流程是设计单位完成设计后，委托效果图制作公司绘制设计最终建成效果的模拟场景效果图。这种与设计过程脱节的效果图所呈现的面目，通常程式化、套路化、标准化，掩盖在纯熟制作技巧下的是对设计背景、立意、目标、思想、形式的不理解。它们通常有着理想化的环境、虚幻的美妙光线、完美的材料质感、色彩，以及完全脱离项目现实人文环境的摩登人、洋人配景人物……当然，也不可否认市场上确实存在一些设计单位，正是假借这种夸张炫目的设计效果图，掩饰设计的缺陷与思维的贫乏，误导委托方对设计品质优劣的真实判断。

　　一直以来，广州土人景观致力于将设计表现融入设计思考的过程中，每位设计师同时也是设计效果图（即市场化称之的"设计效果图"）制作师，注重在磨练每位设计师思考力的同时去锻炼其设计的表现力，要求设计效果图制作的过程与设计思考的过程同样踏实、冷静、诚实。我们利用计算机软件建模或矢量化模拟等手段辅助，认真对工作场地的空间关系、构筑细节、材料、质感、色彩及植物配置进行研究推敲，我们不仅重视效果图最后呈现出的视觉效果，更注重设计表现过程中思维辅助工具的作用，我们把能否准确地表现出项目本身的设计思想、特点、意境作为设计效果图优秀与否的重要标准。反对为了表现而去表现，设计效果图所呈现的最终效果是结果而不是目的。因此，广州土人景观的设计效果图呈现出与公司项目设计风格相吻合的特点。

　　今天，广州土人景观将多年来较为优秀的效果图集结成书，一方面希望借此书的出版弥补当前景观设计效果图书较为缺乏，景观效果图写实不足、写意有余的缺陷；另一方面也希望每位读者能从具体的每张设计效果图、每一个设计表现技法中，看到深植于中国社会现实土壤中的广州土人景观，在拒绝浮躁、平庸、恶俗的立场中，不断磨炼思想力、想像力、创新力，所保持的对景观设计最平实又最新鲜的那份坚持与热爱。

黄征征
国家一级注册建筑师、高级建筑师
广州土人景观顾问有限公司 董事、设计总监
《景观设计》杂志编委
2012 年 3 月 1 日

The design speaks through "pictures" and the concept of design is also conveyed through "pictures". The design rendering is an indispensible tool throughout the design process. Therefore, in the design education and practice, the techniques of rendering become a fundamental skill that all the designers and designer-to-be make up their mind to grasp through repeated drilling. The forms of design rendering are various. It can be the early draft at the beginning of the formation of concepts, or the manuscript conducive to analysis amid the thinking process or the final designing results. The most common form is a simulated rendering of the three-dimensional scenes. Giving a clear glimpse to the non-professionals about design, design rendering of this kind is the most popular and figures prominently in this book. Apart from that, the forms of design rendering also include the vectorized CAD graphical representation, embodying as content the plan and site-plan reflecting the layout positioning relations, the profile map showing the vertical relations, various analysis pictures displaying the design structural relations and elevation drawing showing the aspect effects. These abstract rendering forms are mainly applied to the design rendering of professional technologies.

In China, the making of design rendering has developed into a specialized and marketized industry. Independent of the designers, the drawing of rendering can be carried out by people not specializing in design in a production-line way. The usual work procedure goes as follows: the designing party commissions the rendering-making company to draw the simulated scene rendering after it has completed the design work. Rendering made out of joint with the designing process are usually stylized, ritualized and standardized. Under the cover of the dexterous drawing skill roars the poor understanding of the background, purpose, aim, content and form of the design. Rendering of this kind tends to have the idealized surroundings, surreal lighting, perfect material texture, coloring and the modern and foreign people as background figures totally out of place with the actual human environment of the project. It is not rare that some designers manage to disguise the flaws of the design and the barrenness of their thinking mode with this kind of flashy and flamboyant rendering, thus blinding the entrusting party in judgment of the quality of the design.

Guangzhou Turen Landscape Planning CO, LTD has always been devoted to incorporating the design rendering into the thinking process of design. The designer also acts as the maker of design rendering (namely the so-called "design sketch" during the period of marketization). Special emphasis is put upon the expressive force of the design as well as the thinking power of every designer. And the requirements have been that the process of making the design rendering be as steady, calm and honest as that of design reflection. With the assistance of the model-building using the computer software and vectorized simulation in delving into the spatial relations, structural details, materials, textures, colors and plant layout in the workplace, we attach importance to not only the final visual effects of the rendering but also the roles the auxiliary tools of thinking play in the process of design and rendition. In assessing the design rendering, we pay special attention as to whether it accurately echoes the design objective, features, atmosphere of the project itself. We strongly oppose the practice of rendering for rendering's sake, for the final effects displayed by the design rendering is a means to the end, not the end itself. Therefore, the design rendering of Guangzhou Turen Landscape Planning CO, LTD is characterized by the dovetailing of design rendering with the design style of the corporate project.

Today, Guangzhou Turen Landscape Planning CO, LTD compiles the excellent works over the years into a book, partly to make up for the fact that the landscape design rendering books are relatively scanty and that many of their landscape rendering put excess emphasis upon the amorphous mood and seem wanting in displaying the substantial real life. Besides, I hope in perusing the concrete design rendering and techniques, the readers could identify with Guangzhou Turen Landscape Planning CO, LTD which has been deeply embedded in the soil of China's social reality and temper their thinking power, imagination and creativity while standing their ground of resistance to frivolousness, mundanity and flaunt. In the end, nothing will remain but the purest and freshest persistent love of landscape design.

Zhengzheng Huang
National First-class Registered Architect, Senior Architect
Director and Design Director of Guangzhou Turen Landscape Planning CO, LTD
Member of Editorial Committee of Landscape Design Magazine
March 1st, 2012

INDEX / 索引

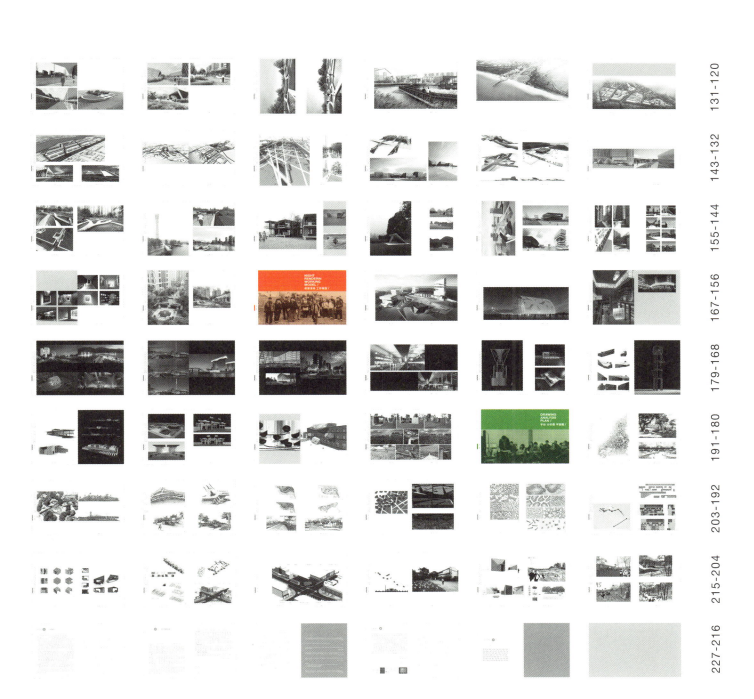

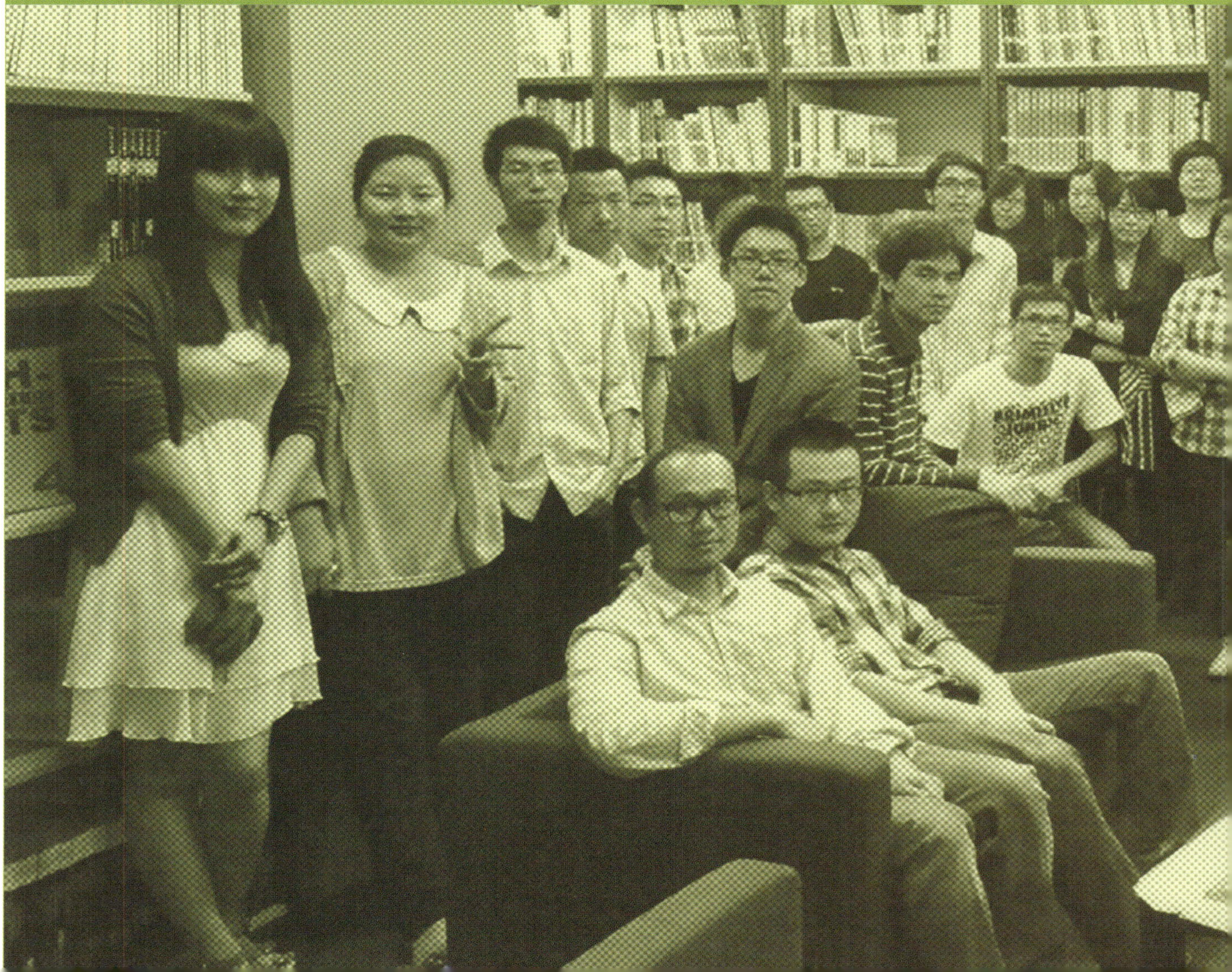

COLLAGE RENDERING /

拼贴合成 /

001 项目地点 / Project Location：广东省广州市 / Guangdong，Guangzhou 设计时间 / Design Time：2010

001 项目地点 / Project Location: 广东省广州市 / Guangzhou, Guangdong 设计时间 / Design Time: 2010

002 项目地点 / Project Location: 广东省四会市 / Guangdong, Sihui 设计时间 / Design Time: 2009

001

热电博物馆
Power Station Museum

002

001

002

003

001 项目地点 / Project Location: 广东省东莞市 / Guangdong, Dongguan 设计时间 / Design Time: 2010
002 项目地点 / Project Location: 广东省东莞市 / Guangdong, Dongguan 设计时间 / Design Time: 2010

001

002

SKETCHUP
RENDERING /
草图模型渲染 /

001 项目地点 / Project Location: 广东省深圳市 / Guangdong, Shenzhen 设计时间 / Design Time: 2011
002 项目地点 / Project Location: 广东省广州市 / Guangdong, Guangzhou 设计时间 / Design Time: 2011
003 项目地点 / Project Location: 广东省深圳市 / Guangdong, Shenzhen 设计时间 / Design Time: 2011

001

GuangZhou / Turenscape

001

002

项目地点 / Project Location: 广东省深圳市 / Guangdong, Shenzhen 设计时间 / Design Time: 2011
项目地点 / Project Location: 广东省东莞市 / Guangdong, Dongguan 设计时间 / Design Time: 2011
项目地点 / Project Location: 广东省深圳市 / Guangdong, Shenzhen 设计时间 / Design Time: 2011

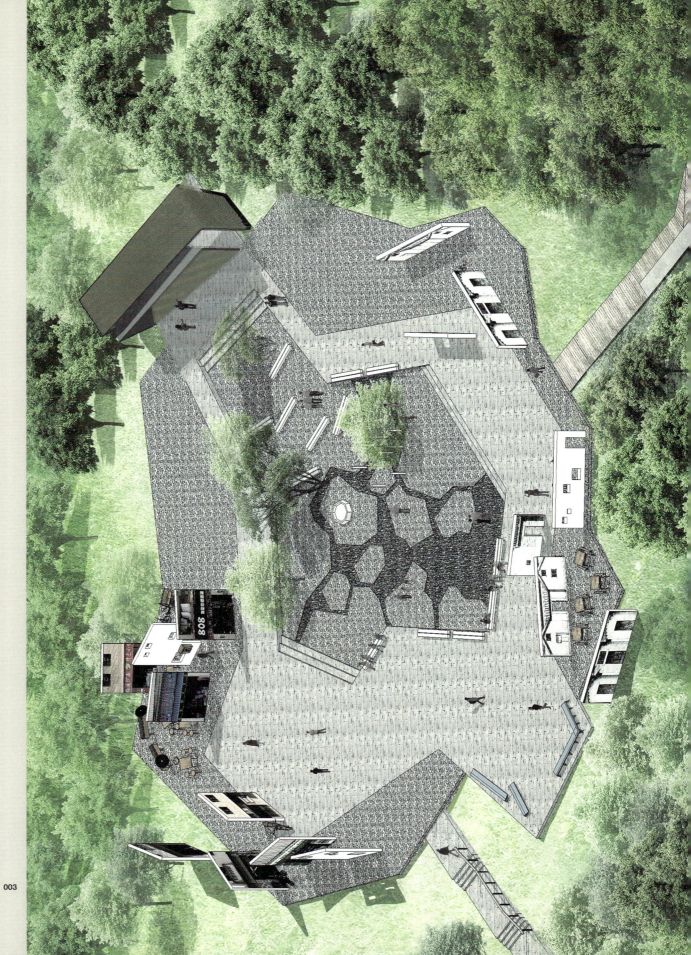

项目地点 / Project Location: 广东省广州市 / Guangdong, Guangzhou 设计时间 / Design Time: 2011

001

001

001 项目地点 / Project Location: 广东省深圳市 / Guangdong，Shenzhen 设计时间 / Design Time: 2011
002 项目地点 / Project Location: 广东省深圳市 / Guangdong，Shenzhen 设计时间 / Design Time: 2011
003 项目地点 / Project Location: 广东省东莞市 / Guangdong，Dongguan 设计时间 / Design Time: 2006
004 项目地点 / Project Location: 广东省深圳市 / Guangdong，Shenzhen 设计时间 / Design Time: 2011

001

002

GuangZhou / Turenscape

001 项目地点 / Project Location: 广东省深圳市 / Guangdong, Shenzhen 设计时间 / Design Time: 2011
002 项目地点 / Project Location: 广东省深圳市 / Guangdong, Shenzhen 设计时间 / Design Time: 2011
003 项目地点 / Project Location: 四川省成都市 / Sichuan, Chengdu 设计时间 / Design Time: 2006

001

002

003

001

项目地点 / Project Location: 广东省深圳市 / Guangdong, Shenzhen　设计时间 / Design Time: 2011
项目地点 / Project Location: 广东省广州市 / Guangdong, Guangzhou　设计时间 / Design Time: 2011
项目地点 / Project Location: 广东省深圳市 / Guangdong, Shenzhen　设计时间 / Design Time: 2011

001
002
003

002

001 项目地点 / Project Location: 四川省成都市 / Sichuan, Chengdu 设计时间 / Design Time: 2006
002 项目地点 / Project Location: 四川省成都市 / Sichuan, Chengdu 设计时间 / Design Time: 2006

001

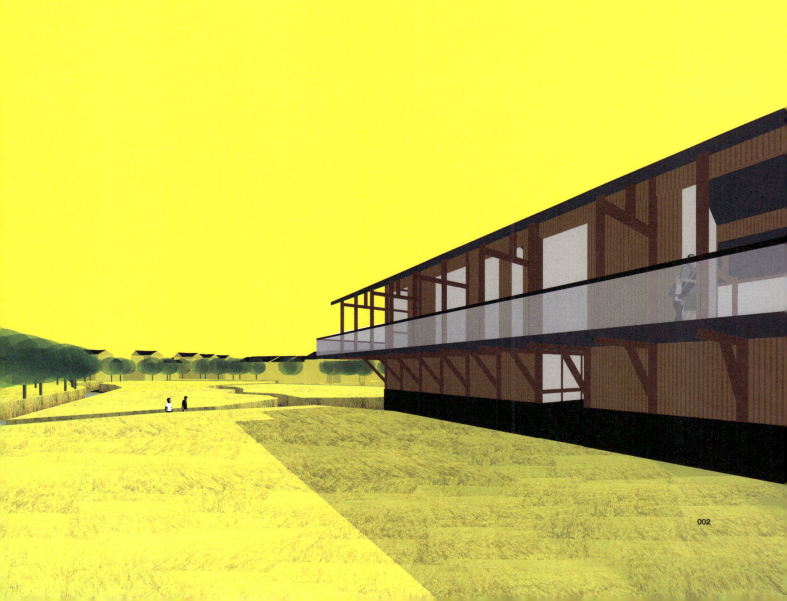

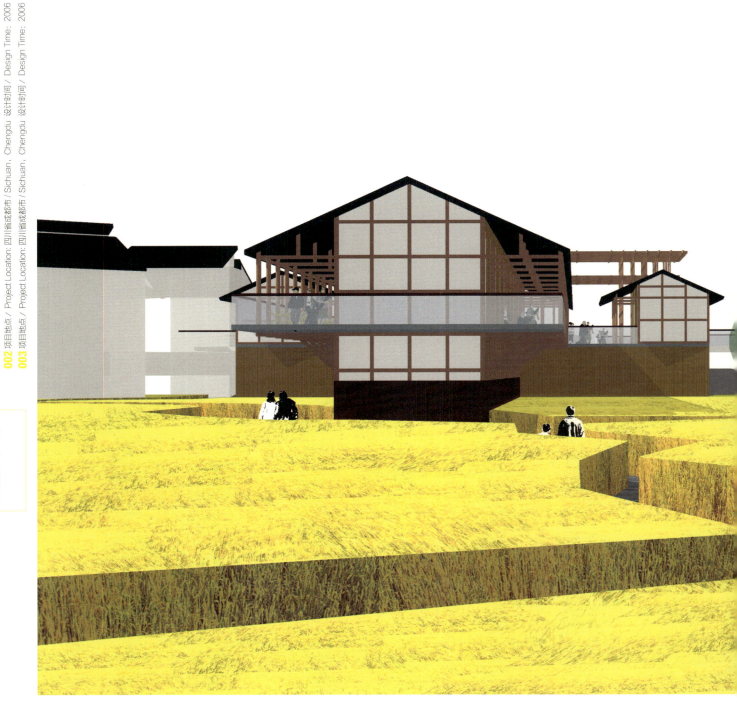

001

002

003

001 项目地点 / Project Location: 广东省东莞市 / Guangdong, Dongguan 设计时间 / Design Time: 2010
002 项目地点 / Project Location: 广东省东莞市 / Guangdong, Dongguan 设计时间 / Design Time: 2009
003 项目地点 / Project Location: 广东省东莞市 / Guangdong, Dongguan 设计时间 / Design Time: 2010

001

001

002

项目地点 / Project Location: 广东省东莞市 / Guangdong, Dongguan 设计时间 / Design Time: 2009
项目地点 / Project Location: 广东省深圳市 / Guangdong, Shenzhen 设计时间 / Design Time: 2006
项目地点 / Project Location: 内蒙古鄂尔多斯市 / Inner Mongolia, Ordos 设计时间 / Design Time: 2009

001
002
003

003

001

002

001 项目地点 / Project Location: 广东省佛山市 / Guangdong, Foshan 设计时间 / Design Time: 2009
002 项目地点 / Project Location: 广东省佛山市 / Guangdong, Foshan 设计时间 / Design Time: 2009
003 项目地点 / Project Location: 广东省佛山市 / Guangdong, Foshan 设计时间 / Design Time: 2009
004 项目地点 / Project Location: 广东省广州市 / Guangzhou, Guangdong 设计时间 / Design Time: 2009

广州土人景观效果图表现十年集

003

004

001 项目地点 / Project Location: 广东省广州市 / Guangdong, Guangzhou 设计时间 / Design Time: 2010
002 项目地点 / Project Location: 广东省深圳市 / Guangdong, Shenzhen 设计时间 / Design Time: 2007
003 项目地点 / Project Location: 广东省东莞市 / Guangdong, Dongguan 设计时间 / Design Time: 2010
004 项目地点 / Project Location: 广东省东莞市 / Guangdong, Dongguan 设计时间 / Design Time: 2009

001

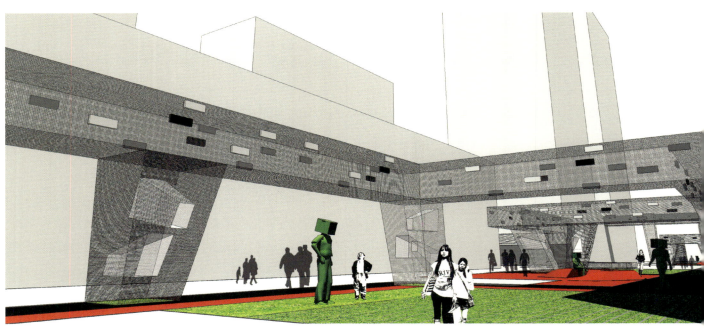

002

003

004

001 项目地点 / Project Location: 广东省东莞市 / Guangdong，Dongguan 设计时间 / Design Time：2010

001

001 项目地点 / Project Location: 广东省深圳市 / Guangdong, Shenzhen 设计时间 / Design Time: 2008
002 项目地点 / Project Location: 广东省东莞市 / Guangdong, Dongguan 设计时间 / Design Time: 2010
003 项目地点 / Project Location: 广东省广州市 / Guangdong, Guangzhou 设计时间 / Design Time: 2010

PAGE / 52

001 项目地点 / Project Location: 广东省东莞市 / Guangdong, Dongguan 设计时间 / Design Time: 2010
002 项目地点 / Project Location: 广东省广州市 / Guangdong, Guangzhou 设计时间 / Design Time: 2011
003 项目地点 / Project Location: 广东省东莞市 / Guangdong, Dongguan 设计时间 / Design Time: 2010

001

001 项目地点 / Project Location: 广东省东莞市 / Guangdong, Dongguan 设计时间 / Design Time: 2010
002 项目地点 / Project Location: 广东省东莞市 / Guangdong, Dongguan 设计时间 / Design Time: 2010

002

001

001 项目地点 / Project Location: 广东省佛山市 / Guangdong, Foshan 设计时间 / Design Time: 2009

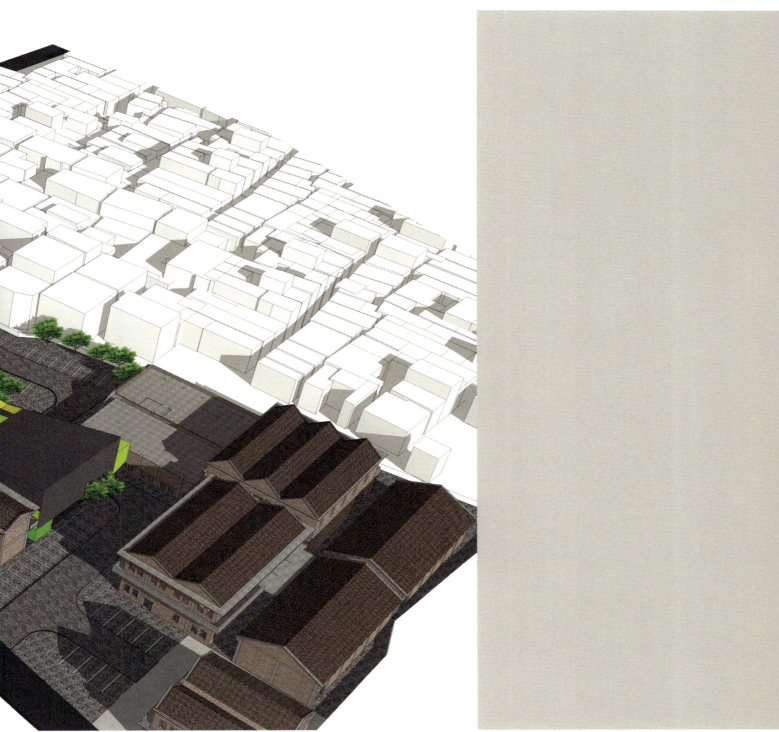

001

001 项目地点 / Project Location: 广东省东莞市 / Guangdong, Dongguan 设计时间 / Design Time: 2010
002 项目地点 / Project Location: 广东省东莞市 / Guangdong, Dongguan 设计时间 / Design Time: 2010
003 项目地点 / Project Location: 广东省东莞市 / Guangdong, Dongguan 设计时间 / Design Time: 2010

001

002

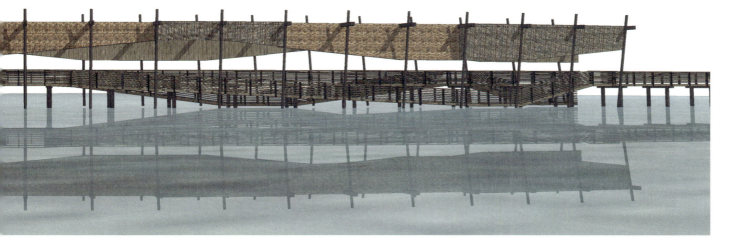

003

001 项目地点 / Project Location: 广东省东莞市 / Guangdong, Dongguan 设计时间 / Design Time: 2010
002 项目地点 / Project Location: 广东省广州市 / Guangdong, Guangzhou 设计时间 / Design Time: 2007
003 项目地点 / Project Location: 广东省广州市 / Guangdong, Guangzhou 设计时间 / Design Time: 2007

001

002

003

001 项目地点 / Project Location: 广东省四会市 / Guangdong, Sihui 设计时间 / Design Time: 2009
002 项目地点 / Project Location: 广东省四会市 / Guangdong, Sihui 设计时间 / Design Time: 2009
003 项目地点 / Project Location: 广东省四会市 / Guangdong, Sihui 设计时间 / Design Time: 2009
004 项目地点 / Project Location: 广东省四会市 / Guangdong, Sihui 设计时间 / Design Time: 2009

002

003

004

项目地点 / Project Location: 广东省佛山市 / Guangdong, Foshan 设计时间 / Design Time: 2009
项目地点 / Project Location: 广东省广州市 / Guangdong, Guangzhou 设计时间 / Design Time: 2007
项目地点 / Project Location: 广东省广州市 / Guangdong, Guangzhou 设计时间 / Design Time: 2007

001

001 项目地点
002 项目地点
003 项目地点

002

003

001 项目地点 / Project Location: 广东省佛山市 / Guangdong, Foshan 设计时间 / Design Time: 2008
002 项目地点 / Project Location: 广东省佛山市 / Guangdong, Foshan 设计时间 / Design Time: 2009
003 项目地点 / Project Location: 广东省深圳市 / Guangdong, Shenzhen 设计时间 / Design Time: 2011
004 项目地点 / Project Location: 广东省阳江市 / Guangdong, Yangjiang 设计时间 / Design Time: 2009

004

001 项目地点 / Project Location: 广东省阳江市 / Guangdong, Yangjiang 设计时间 / Design Time: 2009
002 项目地点 / Project Location: 广东省深圳市 / Guangdong, Shenzhen 设计时间 / Design Time: 2008
003 项目地点 / Project Location: 广东省四会市 / Guangdong, Sihui 设计时间 / Design Time: 2009

PAGE / 70

001

003

001 项目地点 / Project Location: 广东省四会市 / Guangdong, Sihui 设计时间 / Design Time: 2009
002 项目地点 / Project Location: 广东省阳江市 / Guangdong, Yangjiang 设计时间 / Design Time: 2009
003 项目地点 / Project Location: 广东省四会市 / Guangdong, Sihui 设计时间 / Design Time: 2009

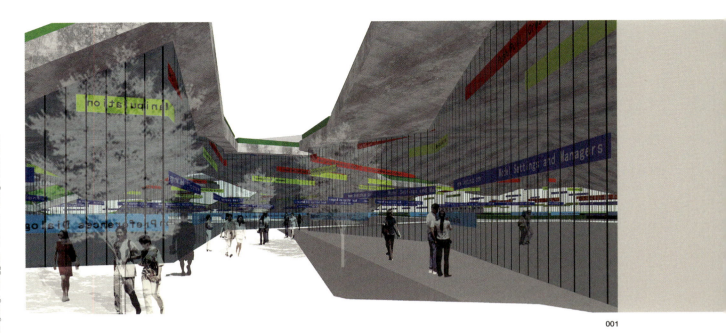

001

001 项目地点 / Project Location: 广东省佛山市 / Guangdong, Foshan 设计时间 / Design Time: 2006
002 项目地点 / Project Location: 广东省东莞市 / Guangdong, Dongguan 设计时间 / Design Time: 2009

001 项目地点 / Project Location: 广东省东莞市 / Guangdong, Dongguan / 设计时间 / Design Time: 2010
002 项目地点 / Project Location: 广东省广州市 / Guangdong, Guangzhou / 设计时间 / Design Time: 2010
003 项目地点 / Project Location: 广东省广州市 / Guangdong, Guangzhou / 设计时间 / Design Time: 2010
004 项目地点 / Project Location: 广东省东莞市 / Guangdong, Dongguan / 设计时间 / Design Time: 2010

001

002

003

004

001 项目地点 / Project Location: 广东省广州市 / Guangdong, Guangzhou 设计时间 / Design Time: 2007
002 项目地点 / Project Location: 广东省广州市 / Guangdong, Guangzhou 设计时间 / Design Time: 2007
003 项目地点 / Project Location: 广东省会市 / Guangdong, Shui 设计时间 / Design Time: 2009
004 项目地点 / Project Location: 广东省会市 / Guangdong, Shui 设计时间 / Design Time: 2009

003

004

001 项目地点 / Project Location: 广东省四会市 / Guangdong, Sihui 设计时间 / Design Time: 2009
002 项目地点 / Project Location: 广东省广州市 / Guangdong, Guangzhou 设计时间 / Design Time: 2010
003 项目地点 / Project Location: 广东省广州市 / Guangdong, Guangzhou 设计时间 / Design Time: 2010

001

001 项目地点 / Project Location: 广东省佛山市 / Guangdong, Foshan 设计时间 / Design Time: 2010
002 项目地点 / Project Location: 广东省佛山市 / Guangdong, Foshan 设计时间 / Design Time: 2010
003 项目地点 / Project Location: 四川省成都市 / Sichuan, Chengdu 设计时间 / Design Time: 2011

001

002

3DSMAX
RENDERING /
3DSMAX 模型渲染 /

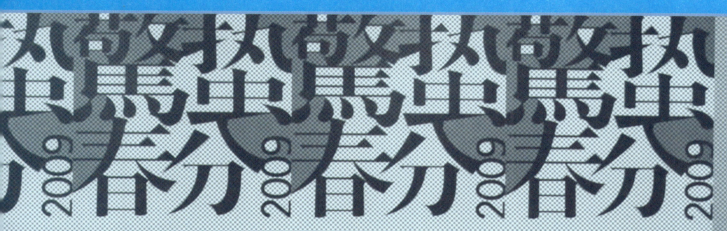

001 项目地点 / Project Location: 广东省中山市 / Guangdong, Zhongshan 设计时间 / Design Time: 2001

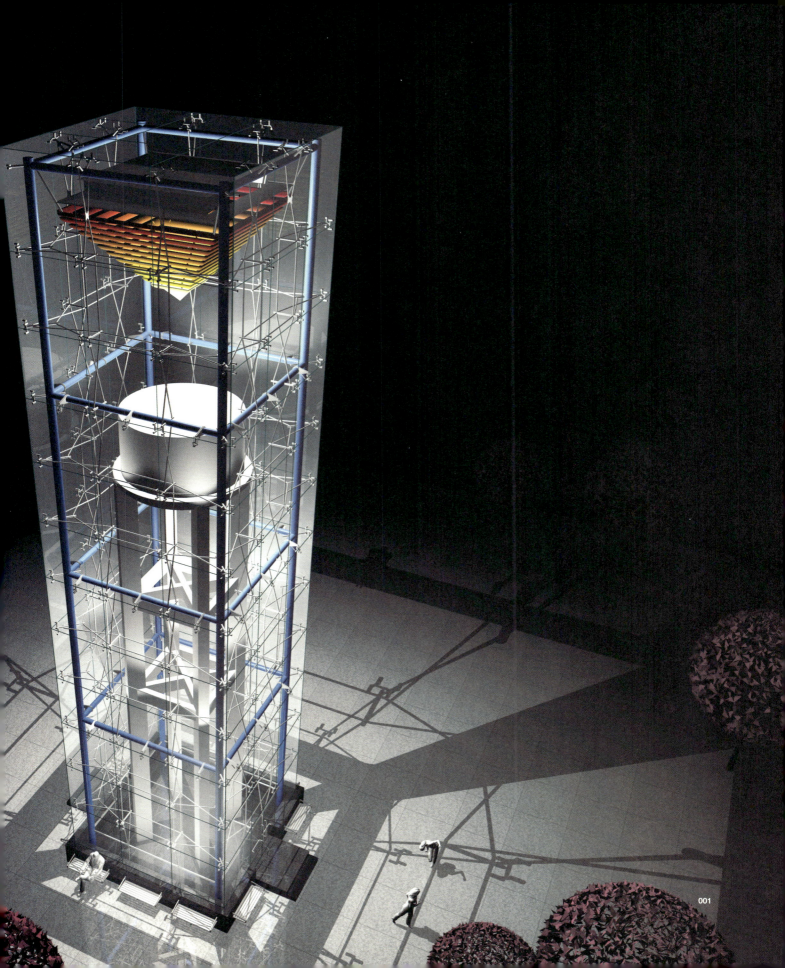

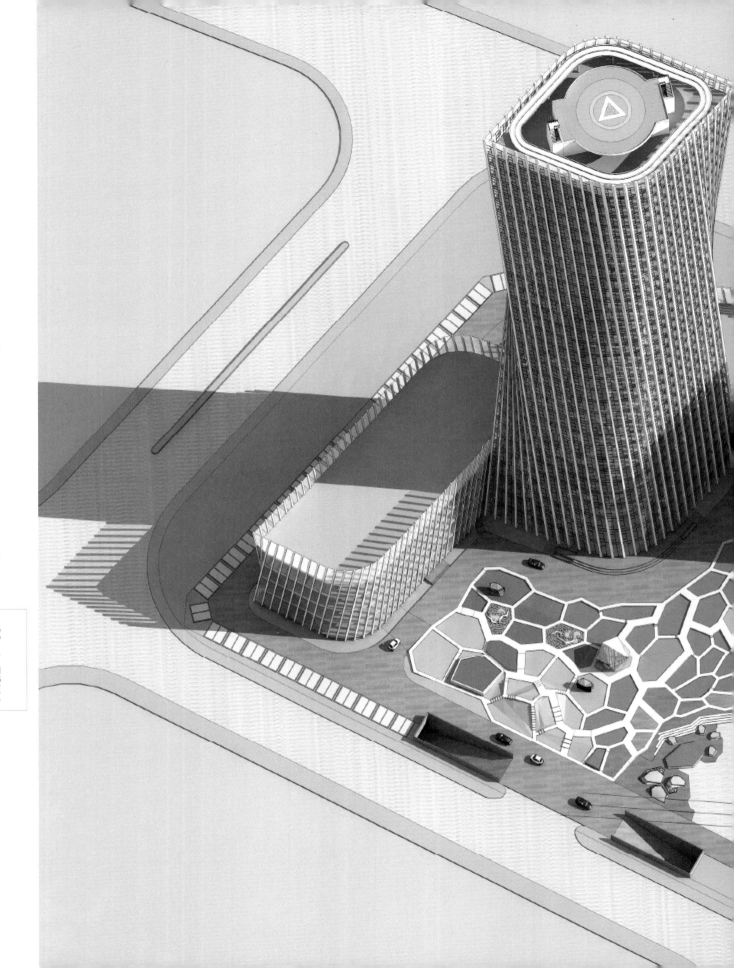

001 项目地点 / Project Location: 广东省佛山市 / Guangdong, Foshan 设计时间 / Design Time: 2008

001 项目地点 / Project Location: 广东省东莞市 / Guangdong, Dongguan 设计时间 / Design Time: 2009
002 项目地点 / Project Location: 广东省东莞市 / Guangdong, Dongguan 设计时间 / Design Time: 2009

001

002

项目地点 / Project Location: 广西省桂林市 / GuangXi，Gulin 设计时间 / Design Time：2004

001 项目地点 / Project Location: 海南省文昌市 / Hainan, Wenchang 设计时间 / Design Time: 2010

001

002

003

惠通桥

001 项目地点 / Project Location: 云南省昆明市 / Yunnan, Kunming 设计时间 / Design Time: 2008
002 项目地点 / Project Location: 江西省南昌市 / Jiangxi, Nanchang 设计时间 / Design Time: 2007

001

001 项目地点 / Project Location : 广东省广州市 / Guangdong，Guangzhou 设计时间 / Design Time：2007
002 项目地点 / Project Location : 广东省广州市 / Guangdong，Guangzhou 设计时间 / Design Time：2007
003 项目地点 / Project Location : 广东省广州市 / Guangdong，Guangzhou 设计时间 / Design Time：2007

001

002

003

001 项目地点 / Project Location: 广东省广州市 / Guangdong，Guangzhou 设计时间 / Design Time: 2007
002 项目地点 / Project Location: 广东省广州市 / Guangdong，Guangzhou 设计时间 / Design Time: 2007
003 项目地点 / Project Location: 广东省广州市 / Guangdong，Guangzhou 设计时间 / Design Time: 2007

001

001 项目地点 / Project Location: 四川省成都市 / Sichuan, Chengdu 设计时间 / Design Time: 2006
002 项目地点 / Project Location: 四川省成都市 / Sichuan, Chengdu 设计时间 / Design Time: 2006

002

001 项目地点 / Project Location: 广东省东莞市 / Guangdong, Dongguan 设计时间 / Design Time: 2010
002 项目地点 / Project Location: 广东省东莞市 / Guangdong, Dongguan 设计时间 / Design Time: 2010

001 项目地点 / Project Location:广东省深圳市 / Guangdong, Shenzhen 设计时间 / Design Time: 2008
002 项目地点 / Project Location:广东省深圳市 / Guangdong, Shenzhen 设计时间 / Design Time: 2008

001 项目地点 / Project Location: 广东省深圳市 / Guangdong, Shenzhen 设计时间 / Design Time：2008
002 项目地点 / Project Location: 广东省深圳市 / Guangdong, Shenzhen 设计时间 / Design Time：2008

PAGE / 116

001

001

001 项目地点 / Project Location: 天津市 / Tianjin 设计时间 / Design Time: 2007
002 项目地点 / Project Location: 天津市 / Tianjin 设计时间 / Design Time: 2007
003 项目地点 / Project Location: 广东省韶关市 / Guangdong, Shaoguan 设计时间 / Design Time: 2004

002

003

001 项目地点 / Project Location: 内蒙古鄂尔多斯市 / Inner Mongolia, Ordos 设计时间 / Design Time: 2009
002 项目地点 / Project Location: 内蒙古鄂尔多斯市 / Inner Mongolia, Ordos 设计时间 / Design Time: 2009
003 项目地点 / Project Location: 内蒙古鄂尔多斯市 / Inner Mongolia, Ordos 设计时间 / Design Time: 2009

001

002

001 项目地点 / Project Location: 广东省佛山市 / Guangdong, Foshan 设计时间 / Design Time: 2008
002 项目地点 / Project Location: 广东省佛山市 / Guangdong, Foshan 设计时间 / Design Time: 2008
003 项目地点 / Project Location: 河南省开封市 / Henan, Kaifeng 设计时间 / Design Time: 2008

001

002

003

001 项目地点 / Project Location: 广东省佛山市 / Guangdong, Foshan 设计时间 / Design Time: 2008
002 项目地点 / Project Location: 广东省佛山市 / Guangdong, Foshan 设计时间 / Design Time: 2008

001

001 项目地点 / Project Location:河南省开封市 / Henan, Kaifeng 设计时间 / Design Time:2008

001 项目地点 / Project Location:海南省文昌市 / Hainan, Wenchang 设计时间 / Design Time: 2010.

001 项目地点 / Project Location: 内蒙古鄂尔多斯市 / Inner Mongolia, Ordos 设计时间 / Design Time: 2009

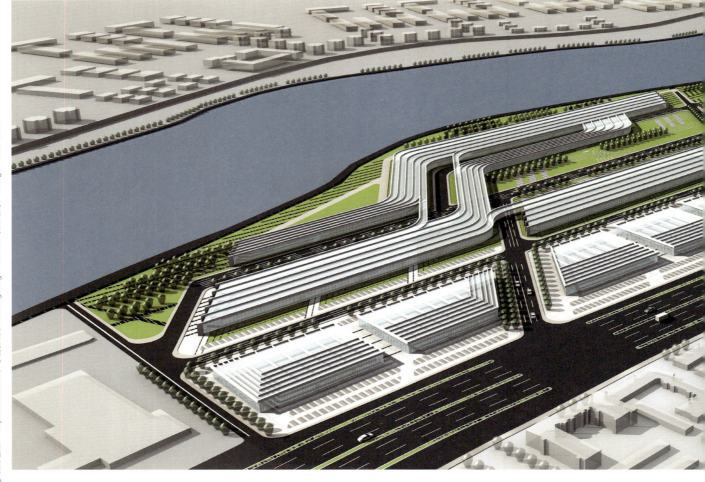

001

003

001 项目地点 / Project Location: 广东省佛山市 Guangdong, Foshan 设计时间 / Design Time: 2006
002 项目地点 / Project Location: 广东省佛山市 Guangdong, Foshan 设计时间 / Design Time: 2006

001

001 项目地点 / Project Location: 广东省广州市 / Guangdong, Guangzhou 设计时间 / Design Time: 2007
002 项目地点 / Project Location: 广东省广州市 / Guangdong, Guangzhou 设计时间 / Design Time: 2007
003 项目地点 / Project Location: 广东省广州市 / Guangdong, Guangzhou 设计时间 / Design Time: 2007

002

001

003

001

001 项目地点 / Project Location: 广东省广州市 / Guangdong, Guangzhou 设计时间 / Design Time: 2006
002 项目地点 / Project Location: 广东省广州市 / Guangdong, Guangzhou 设计时间 / Design Time: 2006
003 项目地点 / Project Location: 广东省广州市 / Guangdong, Guangzhou 设计时间 / Design Time: 2006

PAGE / 138

GuangZhou / Turenscape

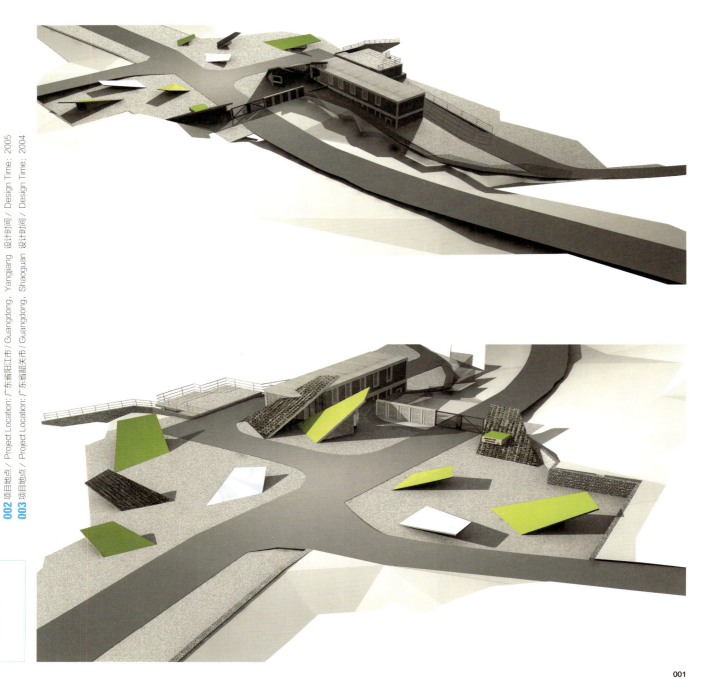

001 项目地点 / Project Location: 广东省韶关市 / Guangdong, Shaoguan 设计时间 / Design Time: 2004
002 项目地点 / Project Location: 广东省阳江市 / Guangdong, Yangjiang 设计时间 / Design Time: 2005
003 项目地点 / Project Location: 广东省韶关市 / Guangdong, Shaoguan 设计时间 / Design Time: 2004

PAGE / 140

001

001 项目地点 / Project Location: 广东省广州市 / Guangdong, Guangzhou 设计时间 / Design Time: 2010
002 项目地点 / Project Location: 江苏省南京市 / Jiangsu, Nanjing 设计时间 / Design Time: 2008

001

001

项目地点 / Project Location: 广东省广州市 / Guangzhou 设计时间 / Design Time: 2005
项目地点 / Project Location: 广东省广州市 / Guangzhou 设计时间 / Design Time: 2004
项目地点 / Project Location: 广东省广州市 / Guangzhou 设计时间 / Design Time: 2005

002

001 项目地点 / Project Location: 广东省广州市 / Guangdong, Guangzhou 设计时间 / Design Time: 2007
002 项目地点 / Project Location: 广东省深圳市 / Guangdong, Shenzhen 设计时间 / Design Time: 2011
003 项目地点 / Project Location: 广东省佛山市 / Guangdong, Foshan 设计时间 / Design Time: 2008

001

002

003

001 项目地点 / Project Location: 广西省桂林市 / GuangXi, Guilin 设计时间 / Design Time: 2004
002 项目地点 / Project Location: 广西省桂林市 / GuangXi, Guilin 设计时间 / Design Time: 2004
003 项目地点 / Project Location: 广西省桂林市 / GuangXi, Guilin 设计时间 / Design Time: 2004

001

002

003

002

003

002

003

001-004 项目地点 / Project Location: 广东省广州市 / Guangdong, Guangzhou 设计时间 / Design Time: 2004
005-012 项目地点 / Project Location: 广东省佛山市 / Guangdong, Foshan 设计时间 / Design Time: 2003

001

002

003

004

005

006

007

008

009

010

011

012

001

002

003

004

广州土人景观效果图表现十年集

005

006

007

008

001 项目地点 / Project Location: 广东省佛山市 / Guangdong, Foshan 设计时间 / Design Time: 2010
002 项目地点 / Project Location: 广东省佛山市 / Guangdong, Foshan 设计时间 / Design Time: 2010

001

002

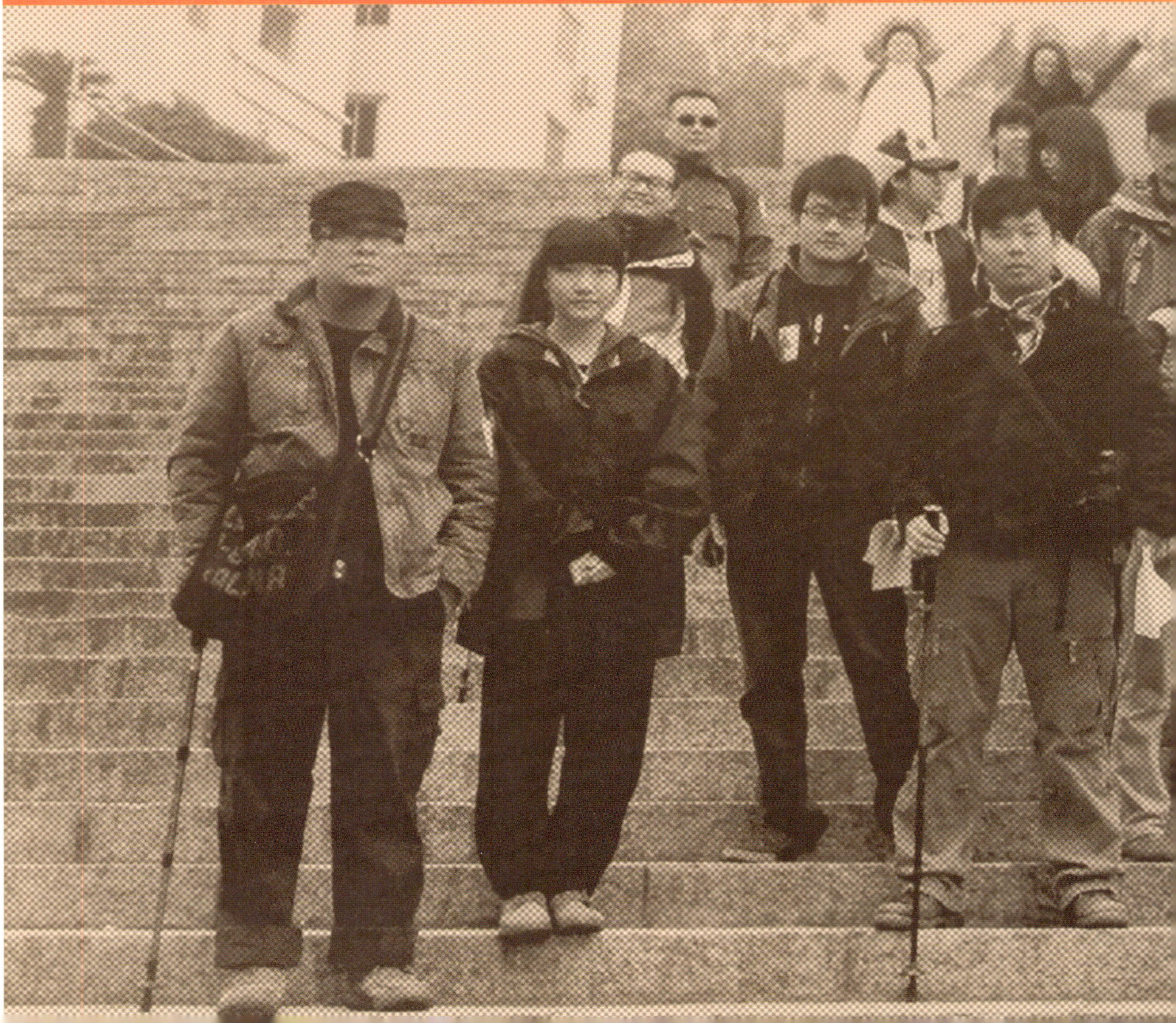

NIGHT
RENDERIN
WORKING
MODEL /

夜景渲染 工作模型 /

001

001 项目地点 / Project Location: 广东省广州市 / Guangdong，Guangzhou 设计时间 / Design Time：2010

热电博物馆
Power Station Museun

A ENTER

Power
Station
Museun

001

001 项目地点：广东省广州市 / Project Location: 广东省广州市 / Guangdong, Guangzhou 设计时间 / Design Time: 2010
002 项目地点：江苏省镇江市 / Project Location: 江苏省镇江市 / Jiangsu, Zhenjiang 设计时间 / Design Time: 2010

GuangZhou / Turenscape

001 项目地点 / Project Location: 广东省东莞市 / Guangdong, Dongguan 设计时间 / Design Time: 2009
002 项目地点 / Project Location: 广东省东莞市 / Guangdong, Dongguan 设计时间 / Design Time: 2010
003 项目地点 / Project Location: 广东省广州市 / Guangdong, Guangzhou 设计时间 / Design Time: 2010
004 项目地点 / Project Location: 广东省广州市 / Guangdong, Guangzhou 设计时间 / Design Time: 2010

001

003

002

004

001

002

003

热电厂

9 号码头

004

001 项目地点 / Project Location: 广东省广州市 / Guangdong, Guangzhou 设计时间 / Design Time: 2009
002 项目地点 / Project Location: 广东省广州市 / Guangdong, Guangzhou 设计时间 / Design Time: 2009
003 项目地点 / Project Location: 河南省开封市 / Henan, Kaifeng 设计时间 / Design Time: 2008

001

002

001 项目地点 / Project Location: 广东省佛山市 / Guangdong, Foshan 设计时间 / Design Time: 2010
002 项目地点 / Project Location: 广东省佛山市 / Guangdong, Foshan 设计时间 / Design Time: 2010

001

002

001 项目地点 / Project Location: 广东省佛山市 / Guangdong, Foshan 设计时间 / Design Time: 2003
002 项目地点 / Project Location: 广东省中山市 / Guangdong, Zhongshan 设计时间 / Design Time: 2001
003 项目地点 / Project Location: 广东省中山市 / Guangdong, Zhongshan 设计时间 / Design Time: 2001

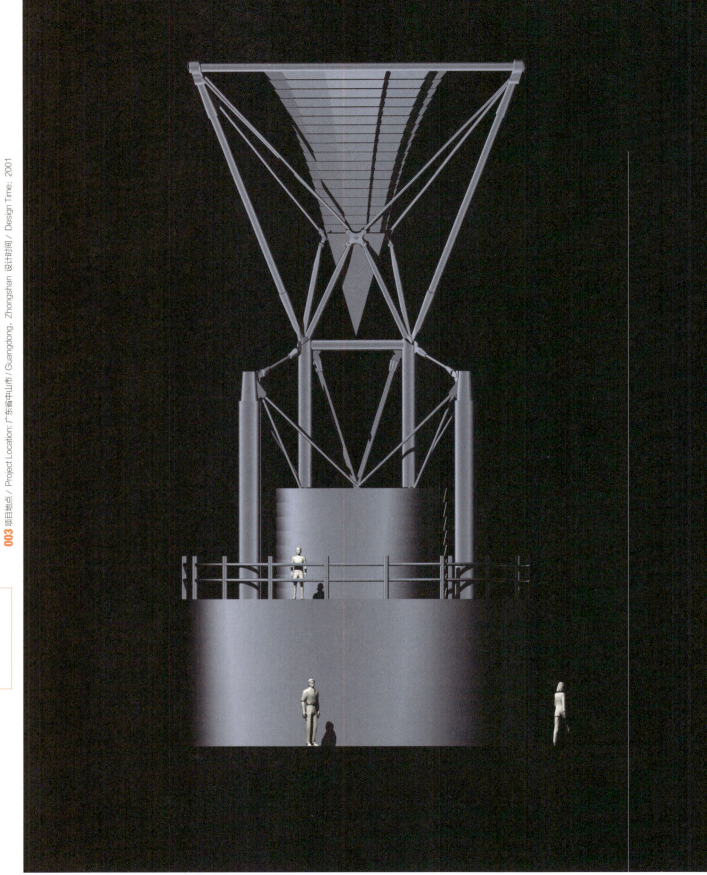

002

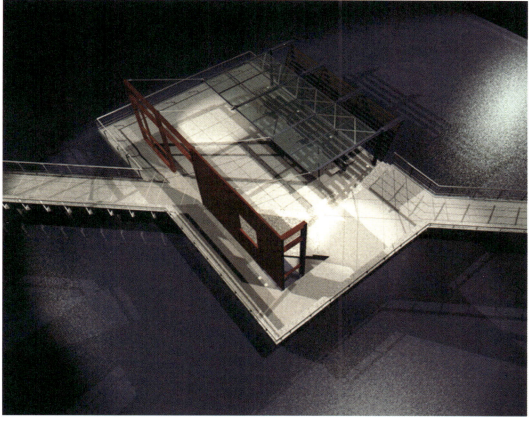

003

001 项目地点 / Project Location: 广东省佛山市 / Guangdong, Foshan 设计时间 / Design Time: 2004
002 项目地点 / Project Location: 广东省中山市 / Guangdong, Zhongshan 设计时间 / Design Time: 2001

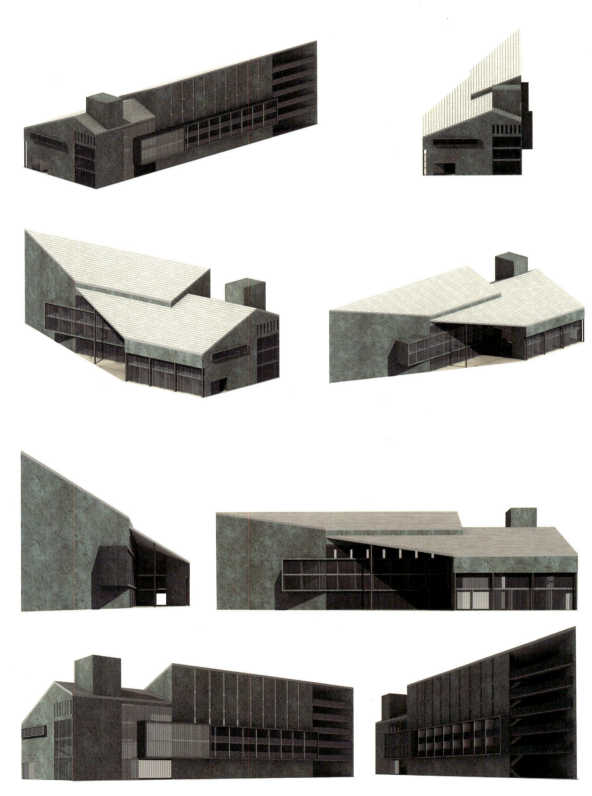

001 项目地点 / Project Location: 广东省佛山市 / Guangdong, Foshan 设计时间 / Design Time: 2003
002 项目地点 / Project Location: 广东省佛山市 / Guangdong, Foshan 设计时间 / Design Time: 2003
003 项目地点 / Project Location: 广东省佛山市 / Guangdong, Foshan 设计时间 / Design Time: 2002
004 项目地点 / Project Location: 广东省佛山市 / Guangdong, Foshan 设计时间 / Design Time: 2002

001

002

东立面

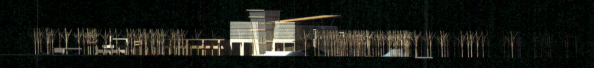

南立面

西立面

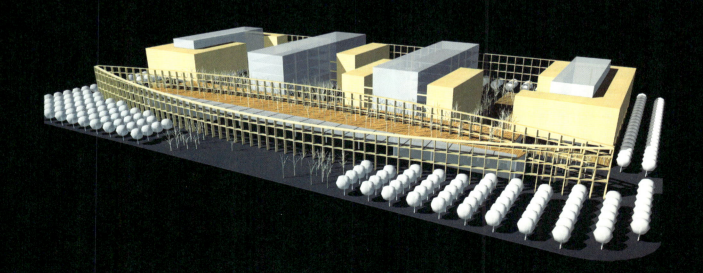

001

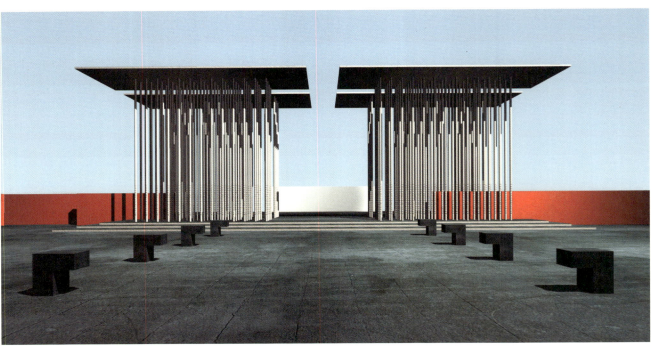

002

001 项目地点 / Project Location: 广东省佛山市 / Guangdong, Foshan 设计时间 / Design Time: 2002
002 项目地点 / Project Location: 广东省云浮市 / Guangdong, Yunfu 设计时间 / Design Time: 2003
003 项目地点 / Project Location: 广东省云浮市 / Guangdong, Yunfu 设计时间 / Design Time: 2003
004 项目地点 / Project Location: 广东省云浮市 / Guangdong, Yunfu 设计时间 / Design Time: 2003

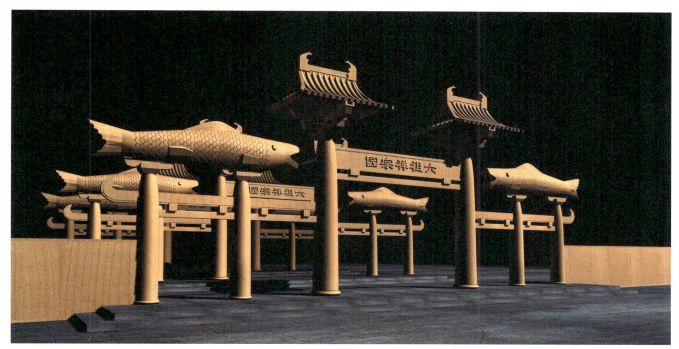

003

004

001 项目地点 / Project Location: 浙江省湖州市 / Zhejiang, Huzhou 设计时间 / Design Time: 2004
002 项目地点 / Project Location: 广东省佛山市 / Guangdong, Foshan 设计时间 / Design Time: 2008

PAGE / 184

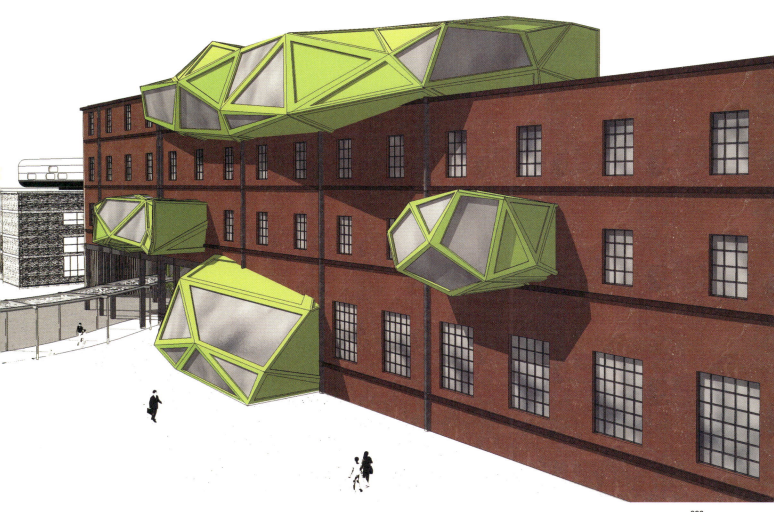

002

001 项目地点 / Project Location: 辽宁省沈阳市 / Liaoning, Shenyang 设计时间 / Design Time: 2003
002 项目地点 / Project Location: 辽宁省沈阳市 / Liaoning, Shenyang 设计时间 / Design Time: 2003

001

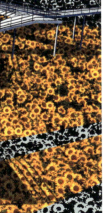

002

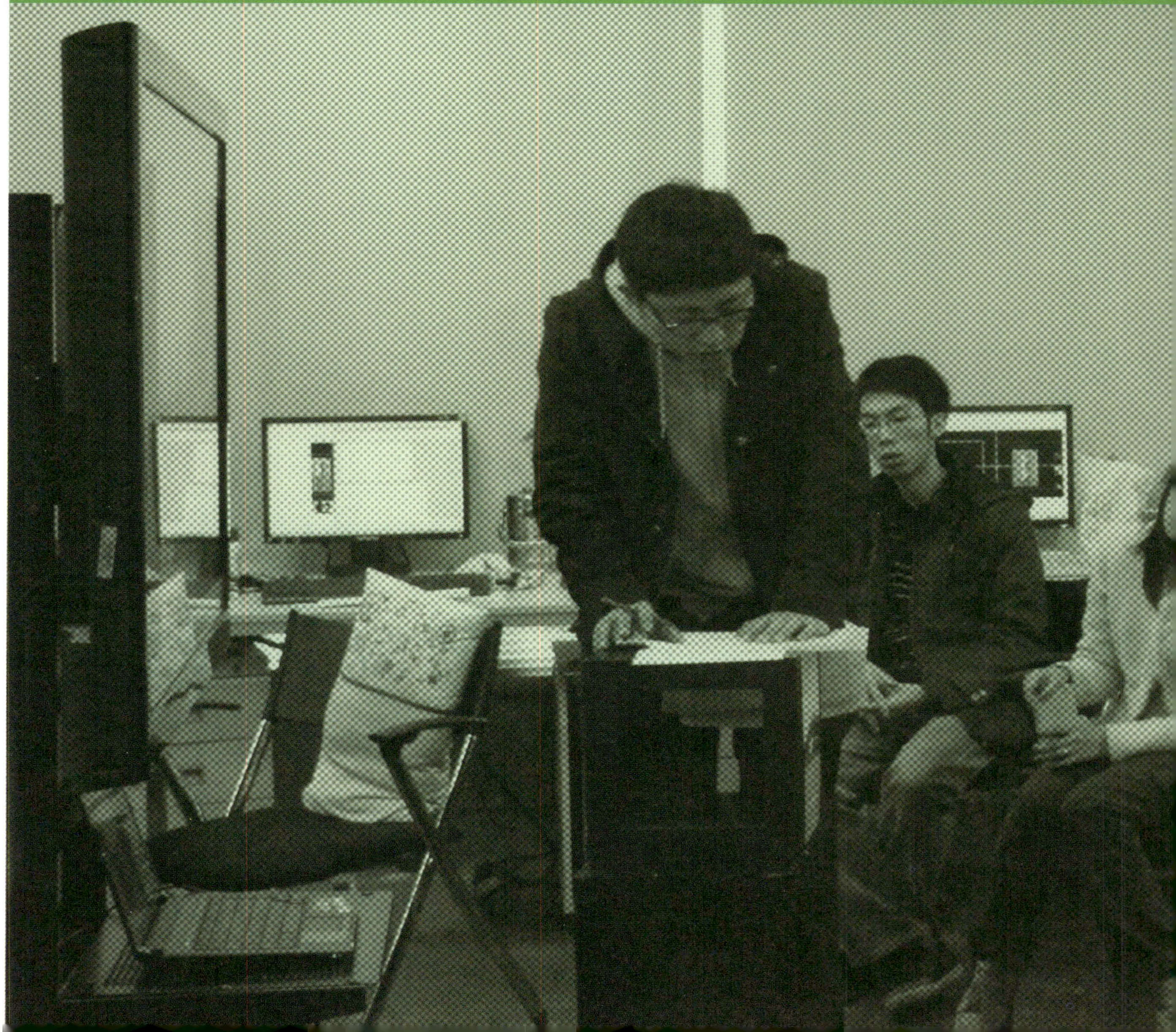

DRAWING
ANALYSIS
PLAN /

手绘 分析图 平面图 /

002

003

001 项目地点／Project Location：广东省广州市／Guangdong, Guangzhou 设计时间／Design Time：2005
002 项目地点／Project Location：广东省珠海市／Guangdong, Zhuhai 设计时间／Design Time：2002
003 项目地点／Project Location：广东省珠海市／Guangdong, Zhuhai 设计时间／Design Time：2002

001

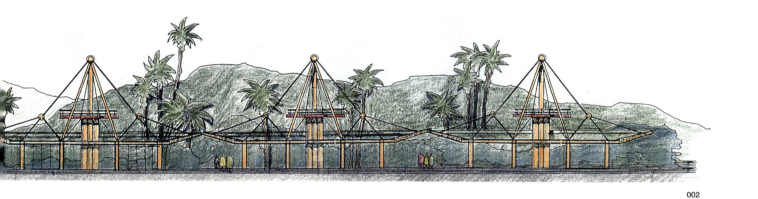

002

003

001 项目地点 / Project Location: 广东省珠海市 / Guangdong, Zhuhai 设计时间 / Design Time: 2002
002 项目地点 / Project Location: 广东省珠海市 / Guangdong, Zhuhai 设计时间 / Design Time: 2002
003 项目地点 / Project Location: 广东省佛山市 / Guangdong, Foshan 设计时间 / Design Time: 2004
004 项目地点 / Project Location: 广东省佛山市 / Guangdong, Foshan 设计时间 / Design Time: 2004

景山路

情侣南路

滨海散步道

珠江口

001

003

昌盛路

情侣南路

珠江口

观海木台

观景平台

林阴广场

桩列

休息廊架

滨海栈道

002

004

001-004 项目地点 / Project Location: 广东省佛山市 / Guangdong, Foshan 设计时间 / Design Time: 2004
005 项目地点 / Project Location: 广东省恩平市 / Guangdong, Enping 设计时间 / Design Time: 2007
006 项目地点 / Project Location: 广东省广州市 / Guangdong, Guangzhou 设计时间 / Design Time: 2007

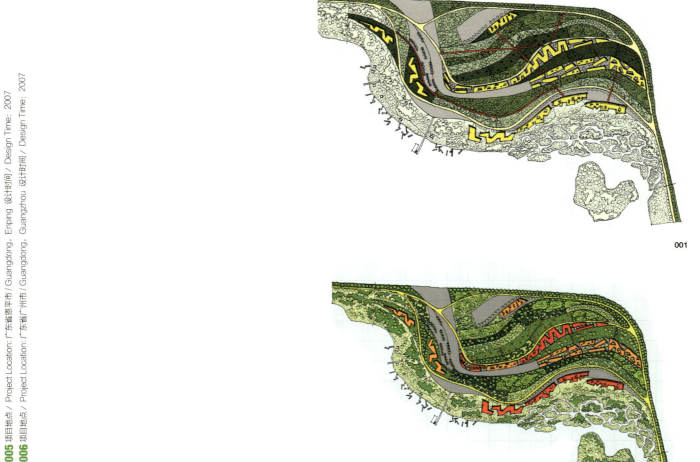

001

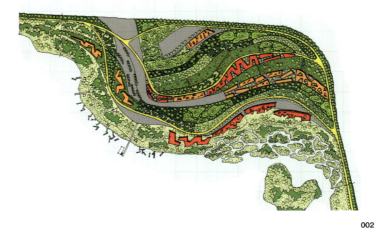

002

005

自然植物群落
林下活动场地
硬质地面
自然植物群落
建筑
机动地块
滨水生态区
儿童活动场地

分区指标
总用地面积：17.592 公顷
滨水生态区：6.35 公顷   占总用地的 36%
中心活动区：6.85 公顷   占总用地的 38.9%
树林群落区：4.39 公顷   占总用地的 25.1%

总平面图

003

004

006

001 项目地点 / Project Location: 广东省佛山市 / Guangdong, Foshan 设计时间 / Design Time: 2004
002 项目地点 / Project Location: 广东省佛山市 / Guangdong, Foshan 设计时间 / Design Time: 2004
003 项目地点 / Project Location: 广东省佛山市 / Guangdong, Foshan 设计时间 / Design Time: 2004

001

002

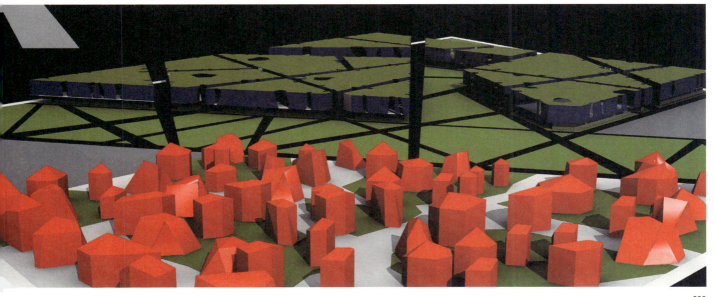

003

001 项目地点 / Project Location: 广东省佛山市 / Guangdong, Foshan 设计时间 / Design Time: 2004
002 项目地点 / Project Location: 广东省佛山市 / Guangdong, Foshan 设计时间 / Design Time: 2004

002

001 项目地点 / Project Location: 广东省广州市 / Guangdong, Guangzhou 设计时间 / Design Time: 2010
002 项目地点 / Project Location: 广东省佛山市 / Guangdong, Foshan 设计时间 / Design Time: 2003
003 项目地点 / Project Location: 广东省佛山市 / Guangdong, Foshan 设计时间 / Design Time: 2003
004 项目地点 / Project Location: 广东省佛山市 / Guangdong, Foshan 设计时间 / Design Time: 2003

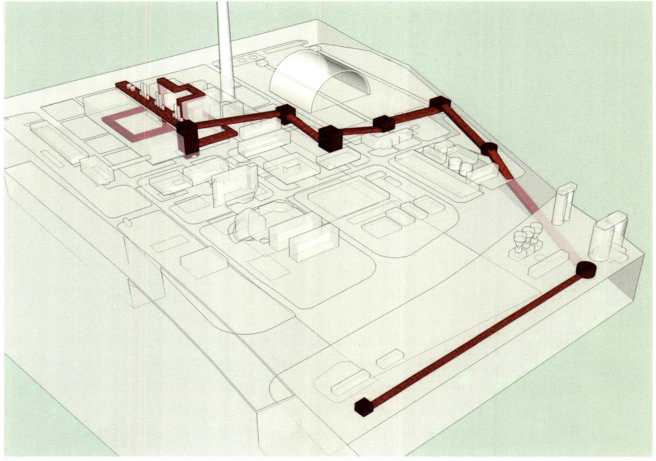

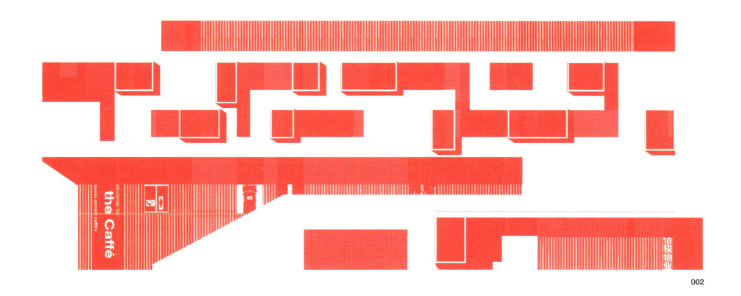

002

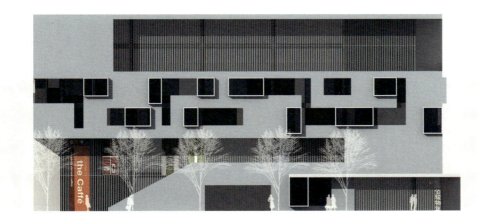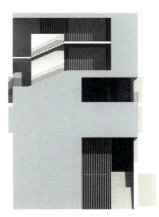

003

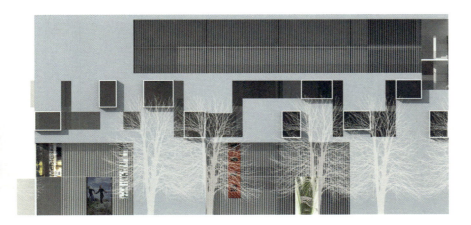

004

001 项目地点 / Project Location: 四川省成都市 / Sichuan, Chengdu 设计时间 / Design Time: 2006
002 项目地点 / Project Location: 广东省佛山市 / Guangdong, Foshan 设计时间 / Design Time: 2004

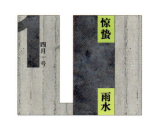

001 项目地点 / Project Location: 广东省广州市 / Guangdong, Guangzhou 设计时间 / Design Time: 2007
002 项目地点 / Project Location: 广东省广州市 / Guangdong, Guangzhou 设计时间 / Design Time: 2007
003 项目地点 / Project Location: 广东省佛山市 / Guangdong, Foshan 设计时间 / Design Time: 2004
004 项目地点 / Project Location: 广东省东莞市 / Guangdong, Dongguan 设计时间 / Design Time: 2010

PAGE / 206

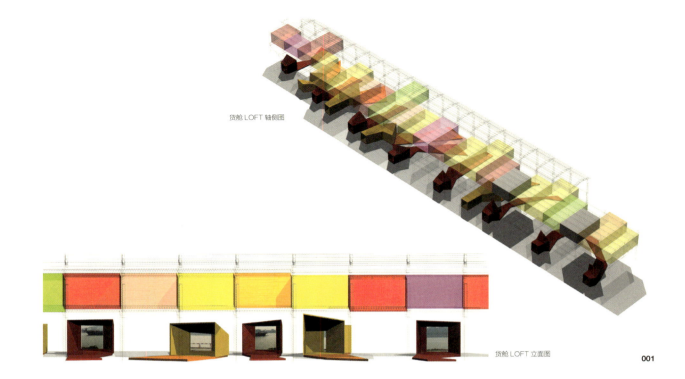

货舱 LOFT 轴侧图

货舱 LOFT 立面图

001

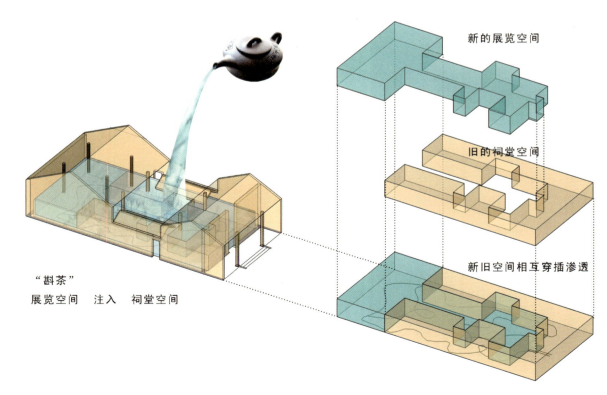

"斟茶"
展览空间　注入　祠堂空间

新的展览空间

旧的祠堂空间

新旧空间相互穿插渗透

003

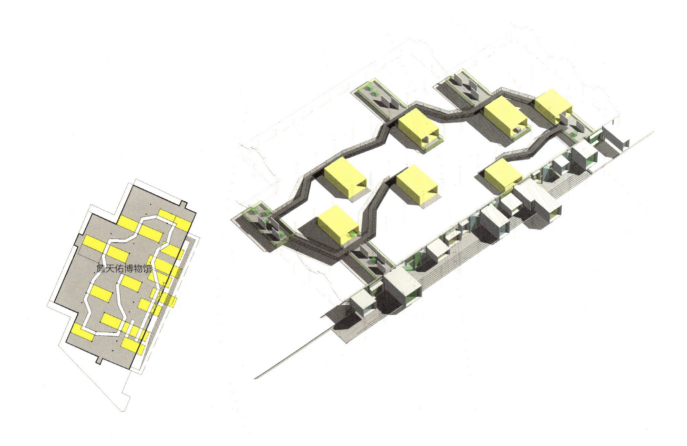

詹天佑博物馆

原状保留　建筑立面改造　建筑立面改造　建筑立面改造　增加绿岛　广告设计　原状保留

建筑立面改造　原状保留　加设路灯设计　大型造型路灯设计

原状保留

增加绿化

建筑立面改造

广告设计

建筑立面改

建筑立面改造　　　　　　　广告设计　　　　　　　建筑立面改造　　广告设计

广告设计　　　　　　大型造型路灯设计　　　　加设路灯设计

001

001 项目地点 / Project Location: 四川省成都市 / Sichuan, Chengdu 设计时间 / Design Time: 2006
002 项目地点 / Project Location: 广东省广州市 / Guangdong, Guangzhou 设计时间 / Design Time: 2006

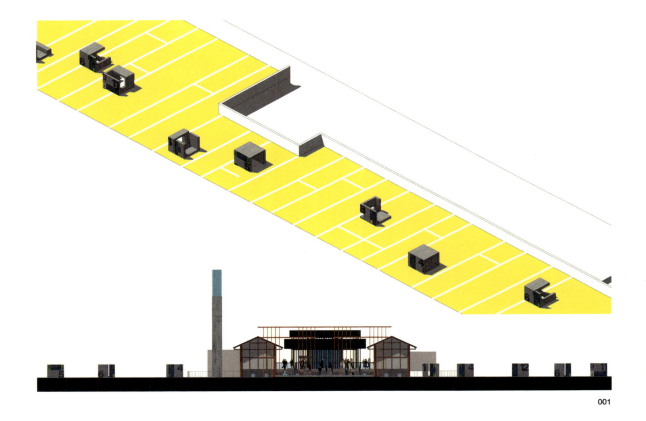

001

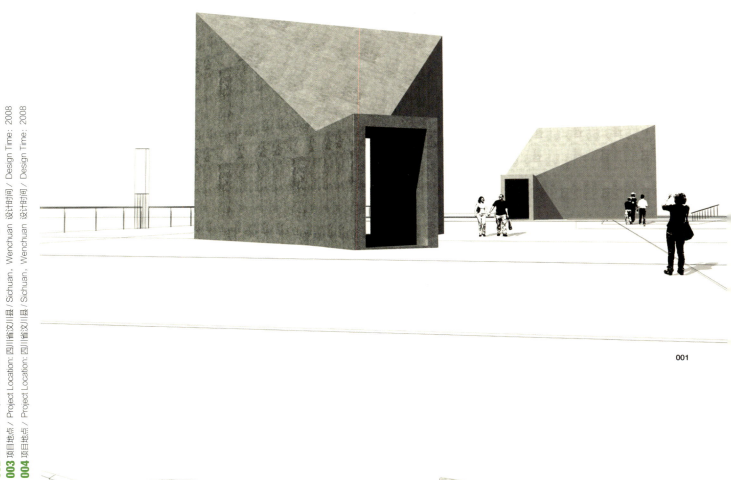

001

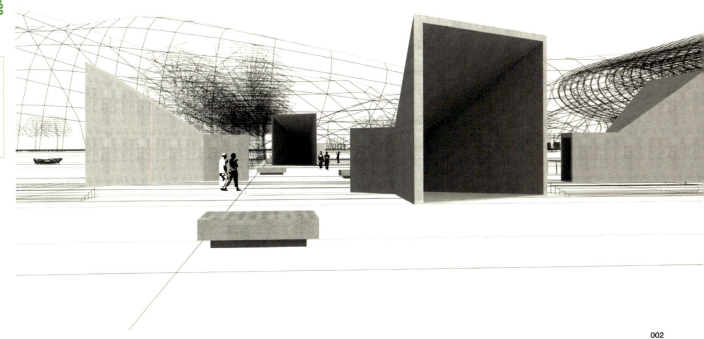

002

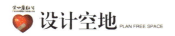

设计空地 PLAN FREE SPACE

游戏空地
避难空地
集会空地
空地的可能性 …

amenity free space
shelter free space
assemblage free space
possibility of hollow…

003

大自然课堂

人类最早的教室

+

现代人的教室

+

2008.5.12

　　大地震后，孩子们的心灵受到了创伤，我们试着去从孩子们的角度考虑，营造一个能让孩子有安全感的环境，让孩子与大自然互动的课堂，用大自然的神奇力量来安抚孩子们心灵。
　　大自然环境具有安抚的作用，能够治愈人心灵的创伤。新鲜的空气、温暖的阳光，以及迷人的花香，都能让孩子们放松心情，变得开朗。
　　儿童天性活泼，心理治疗的最佳方式就是顺应天性，让他们感到快乐，忘记痛苦。户外课堂这种更为自由和灵活的教学方式，更能提高孩子们的学习热情和积极性，并且从中理解生命的意义。

004

003

004

效果图困局
Rendering Dilemma

　　由景观设计公司来出以电脑效果图为主的册子，这在多数人看来是很难想像的。反之，如果这些效果图是手绘的，就会让人觉得很自然。设计公司做方案，效果图公司做效果图，是约定俗成的社会分工。景观公司自己去做大量的电脑效果图，会不会让人感觉不务正业、颠倒主次？抑或是广州土人有强大的效果图团队？也许看完这本图册后，会让您找到自己的答案。

　　早在景观设计这个名词在中国流行起来之前，国内的效果图行业已经依附着以建筑设计和室内设计为主的无数设计公司，逐渐发展起来了。在这个表现设计工作的环节里，效果图公司为广大设计企业提供了非常专业、甚至有些过度的服务。而多数景观设计公司包括景观设计公司无一例外的选择了以手绘为主的表现手法；当然，如今 SketchUp 的泛滥使用也成为了另一种主要表现方法。

　　广州土人从来没有委托其他效果图公司制作效果图，也没有让美丽的手绘图占满方案册，而是采用了比较多元的手法来对设计加以表现。广州土人的效果图很大一部分是直接出自设计师之手。我们也用 3D MAX，也用 SketchUp 和 Photoshop，当然也包括手绘；但是呈现出来的东西却和多数景观公司不同。广州土人不刻意写实、不随意幻想，固执的希望现实的人能参与到虚拟的设计中去，甚至希望每张图都得把设计立场展现出来，就算不能实现，也要当作一个目标摆在前面。我们费尽心思，希望客户能了解我们的想法。与此同时我们的效果图却经常受到客户的冷眼，我们的方案表现曾经被认为是粗糙、平淡、不真实、不富丽堂皇、不国际化的。因此，我们有理由相信有些项目失败的原因就是出于客户对效果图的失望。也许客户需要的是比迪斯尼还艳丽的色彩、比邦德还跌宕的情节，或者是大展宏图的气势，这本身也反映出建设方或者投资方对设计的期待。但是，现实如何？土地、山水、树木和人，这些都不是配景，而是真实存在的，更是这里的主景。

　　效果图永远也无法确保设计就是未来真实的样子（效果图设计的初衷只不过是方便设计方与非设计方的沟通，以便其更好、更真实、更准确的理解设计意图）。但不知从何时起，大家已丧失了现实感，或者说对现实的兴趣；只有令人热血沸腾的效果图才能支撑起他们接受方案的决心。到底什么样的效果图才能发挥它本应具有的说服力？这样的问题曾不止一次的问过自己——美国高线公园（High Line Park）因为效果图太酷，曾被部分纽约人质疑；扎哈•哈迪德（Zaha Hadid）因效果图的出色被人称为"纸上建筑师"；无数中国楼盘的效果图也早已成为了民众心目中美丽的谎言。但是，一张图决定一个项目、一段动画赢得一个投标的故事仍然在上演。效果图是否要服从客户的喜好，是效果图制作中经常遇到的问题。其实，我们也一直在寻求平衡。我们不想把自己摆在边缘人的位置，也不愿当"纸上设计师"，更不应该抛弃表达设计的主动权。从某种角度，我宁愿相信没有效果图这个行业。假设有一天，不用再为效果图伤透脑筋，客户也不再抱着效果图谈设计；方案就是方案，只有平面、立面、剖面、透视，不再有"效果图"这个名称……当然，这只是假设！但是，把虚假的效果图作为推销设计的核心内容是不是对设计的伤害？或者是对客户的不负责任呢？

　　大家一定见识过市面上某些效果图的如梦如幻、恢弘磅礴，其中一些画面效果已经不亚于好莱坞的电影场景。但这些让我最先想到的是当下那些陈词滥调、漫天飞舞的口号，而广州土人的设计是很难与一些过份夸大、不实际的尊贵和雍容扯上半点关系的。效果图必然要围绕着设计来展开，也要和方案设计工作一样理解和尊重设计。广州土人的效果图是没有固定不变的原则的，一套方案里面会有多种不同的做法。前面提到手绘，广州土人的设计师群体里并不缺少手绘能力很棒的人，可是最终用手绘表现的项目并不多。因为设计效果图不是风景画，工具也是多样的，熟练使用它们是工作的前提。为设计做效果图，首先要知道的是做什么，为什么做，然后才是怎么去做。

　　广州土人因为有自己的坚持才会有今天，效果图设计也是一样，没有什么是轻轻松松的。为了本书的出版，本人重新翻查了过去近十年的工作图册，最后发现这本书所收录的图片大概是所能找到效果图的十分之一。大量的图将继续留在电脑里，给现在和将来的广州土人参考和批评。这里面有精彩也有失败，其中的困惑、艰辛，想必每一位广州土人的设计师都有体会。借此书的出版，真心感谢与我们一同战斗过的兄弟姐妹。

黄志坚
广州土人景观顾问有限公司 制作室主任

To most architects, it is awkward that a book which mainly consists of computerized renderings was an effort of a landscape design company, but more natural if the drawings were hand-sketched. It is traditional social labor segmentation that the planning company in charge of the design while the specialized image company in charge of the renderings. Guangzhou Turen Landscape Planning CO, LTD generates most of the renderings in this book. Is it because the company pays too much attention on the business which is not their principal work? Or does it possess a great image proceeding team? You will get the answer after you read this book.

Long before landscape design had its name pop up in China, the domestic rendering industry has developed in dependence upon millions of design companies focused on architectural or interior design. In this presenting sector, specialized image companies tend to provide very professional and even exaggerating services. However, without exception, nearly all the landscape design companies choose hand-sketching to present the project. Nowadays, the SketchUp becomes another major presenting method due to its rampant use.

Guangzhou Turen Landscape Planning CO, LTD has never entrusted a company to provide renderings nor fill the blueprints with fancy hand drawings. We applied a comprehensive way to present our design. Most of the renderings are efforts of the designers. We use 3DMAX, SketchUp, Photoshop, and also, hand-sketching, but the results are so different from other company's works. We never try to stick to realism or fantasy, we keep it in mind that all people who live in the real world could participate in the virtual one, and every sketch to show the panorama of the designer, even if it is just an eternal destination that we are pursuing to. We rack our brains to make our clients understand our idea and concepts. However, sometimes we eat mutton cold from the clients because they are disappointed with our "rough, plain and inveracious" renderings. There is a good reason to believe that some projects failed because the clients were disappointed with the renderings. Perhaps what they need is the colors that are more gorgeous than Disney's Magic Kingdom, plots that are more unconstrained than 007 movies and much more magnificent momentum. However, the main character is the land, the landscape, the trees and human, they are the center piece.

Renderings could never describe the future for sure. (The renderings are design to make the communication between the designers and the clients more convenient, and help them understand the idea of the design.) But at some point, people lost their interest on the reality and became zealous for fabulous pictures which are so unrealistic. I have asked myself for thousands of times, what kind of renderings can play the role it does? High Line Public Park has faced too many doubts form the New York citizen because the beautiful effect image is beyond the reality. Zaha Hadid, who's excellent for great renderings has been called a "paper architect". Countless Chinese real estates renderings have become a beautiful lie for the people, but the story of winning a bid with a rendering or an amazing animation stays alive. Should we compromise and make the rendering according to the clients' favor? We have been searching a path to solve this dilema. There is no doubt that we never mean to be a marginal man, nor become a "paper architect". Sometimes, I would rather there were no rendering business or stuff called rendering. Our clients would never stick to the picture when talking about the design with us. There would be the planning, plane, vertical surface, profile surface and perspectives, but no rendering. Of course, this is just my fantasy. But making an unrealistic rendering the main course of marketing shows no respect to the designers, nor responsibility to the clients.

You must have seen some dreamlike and tremendous renderings, some of which can even rival the scene in Hollywood movies. But it reminds me of the clichés and slogans at the first time, while the design of Guangzhou Turen has no relationship with the exaggerated and impractical prosper. Renderings have to be made on basis of the planning, to show understanding and respect for the design as what the scheme design does. There is no fixed processing principle for the renderings, so you may see various ways of dealing with the renderings. We have a number of great sketchers, but we never rely on sketching and merely have few renderings sketched. Renderings are not landscape paintings, skilled in using renderings is the precondition of the work. You have to know what you want, why you make it and then how to make it before you get start.

Guangzhou Turen struggled its way to the present scale and achievements. In order to publish this book, I have looked through our work pictures of the past 10 years, and find the part we publish is only 10 percents of the existing image stock. The rest pictures will be recorded in computer for further reference and criticism from colleagues of Turen. I want to take this opportunity to thank everyone who have fought together with us and wish every planning could be carried out successfully.

Zhijian Huang

Chief Director of Workshop of Guangzhou Turen Landscape Planning CO, LTD

我和效果图的故事

The Story of Rendering and Me

身为广州土人的设计师，很荣幸能参与到本书的编写。接下来所要说的是六年来我在效果图制作过程中的一些故事。

危机时刻

每次做项目所要面对的困难都是不同的，能够轻轻松松完成的项目至今还没碰到过。危机总是在你意想不到的情况下出现，记得在做东莞某个旅游项目的时候就碰到了危机——在交图的前夕，我们需要做一些能够感动人的效果图，但一张怎样的图才能感动到甲方，没人知道该怎么做。这个时候，庞老师提出了使用超长画幅的方式出图，使图面达到震撼人心的效果，就如同 IMAX 电影那样的超宽屏幕。但是，这么长的画幅如果使用渲染的方法会相当耗费时间。于是，我们立即改成使用 SketchUp 来进行快速渲染，同时配合大量的后期处理，达到类似于电影《阿凡达》那样的梦幻效果，最重要的是这些效果图能够在较短的时间里完成。这些效果图带有一种非现实的梦幻与震撼，同时在图中融合我们的设计，最后得到了甲方的认可，从这时候起也确定了会在一些重要的项目中使用 SketchUp。在这个旅游项目中，效果图并不只是为了表现设计而已，更多的是为了达到一种氛围而制作出来。

不同风格

面对不同的项目会制作出不同风格的效果图，这也正是本书效果图风格各异的原因。我们最初的效果图主要是 3D MAX 渲染为主，但这种方法效率很低，材质和灯光的设置需要花费很多时间反复调整。在渲染完成后，虽然部分图纸可以直接出图，但风格还是偏向于概念图，要有更好的效果则依然需要大量的后期处理。毫无疑问这些 3D MAX 效果图能达到十分好的效果，但需要花费大量的时间和精力，所以只在一些关键项目中才会使用这种方法来出图。

另一种则是使用 SketchUp 制作效果图。最初，这种图纯粹是为了快速制作而出现，虽然它能很直接地交代设计内容，但画面表现得是相当简陋，所以只有在万不得已的情况下才会使用（当然，并不是全部的图都会显得效果很差，在成都青城山项目和广州电视塔投标项目的图中就使用了 SketchUp 渲染的风格，出来的效果却充满了平面设计的构成感）。我们将两种风格的优点结合起来，形成了第三种带剪贴风格的图——即在 SketchUp 渲染的图片上进行大量后期处理，使其达到 3D MAX 渲染了 70~80% 的效果，这样节省了不少的时间。最近正在大量使用这种作图方式，既保证了质量也提高了效率。

最近，我们又开始了新的尝试，就是将真实的场景与设计内容相结合：其实，这种做法国外很早就已经做了，但我们却刚刚开始接触，在广州北岸公园、东莞大岭山、东莞石龙街道改造等项目中就采取了这种做法。其优点是直观与方便，马上可以看到设计跟实际场景相结合的效果。但是，这种做法对作图要求很高，甚至在勘查现场拍摄照片时就要对设计内容的表现进行准备，例如在效果图中如何利用现场拍摄到的照片等因素都要进行思考。在制作过程中，这些与设计相结合并作为背景的照片不是立即就能使用，需要进行筛选和调整。要尽量避免将情况反映得太写实或太不现实，对不同表现的要求恰到好处地处理就是最考验能力的地方。倘若处理不好会使重点不突出，也会让整张图显得杂乱无章。

其实，效果图的做法、风格可以是多种多样的，将来也会出现更多的风格；不仅仅是我们的，也可能是其他同行的。

更加生动的表达

在做图的时候，很怕效果图出来的效果死气沉沉；出现这种情况归根结底是因为设计内容和设计思想无法融入效果图中，生搬硬套的去表现。如何表现得更加生动呢？为了达到这个目的，曾尝试采用不同的方法。在大岭山项目中，在勘察现场时拍摄了不少当地居民生活的景象，比如在水里嬉戏的小孩、悠哉的老人、正在犁地的水牛等，这些都是场地原有的画面。当我们将这些生动的场景放入效果图时，感觉仿佛将整个现场在图中展现，甚至设计内容如同建好般跃然纸上。有时候，我们也喜欢在图里开一些小玩笑，让效果图带几分幽默，例如在图中放进眼下的话题人物，甚至是直接将参与设计的同事也偷偷地作为配景人物加到背景中。

在另外的一些项目中，我们则会改变效果图的构图方式，前面所提到的超长画幅是其中一种；或者会对画面进行裁剪或留白，将配景人物放在留白与有实际表现之间的位置上，让人物看上去就像从图外走进图中，使其看上去更加生动。

在一些需要表现有特殊节庆活动的场景时，就需要用到一些特有的素材，例如舞狮、龙舟、跳飞机、打陀螺等，这些元素不一定能在素材库里找到，也会从网络中寻找对应的图片进行加工。但相对于网络，我们更倾向于出去走走、拍拍照，并将拍摄到的内容作为素材。因为有了这些照片，才让一些很沉闷的场景变得有活力。

效果图的不同作用

效果图仅仅是作为方案给甲方看的图纸么？答案显然是否定的。在深圳中科研发园这个项目中，我们并没有做很多效果很好的效果图，但意外的却是甲方对一套能用于施工指导的效果图产生了深刻的印象。当时项目已经进入实际施工阶段，不过由于某个节点的结构复杂，单凭平面图或者剖面图很难令施工人员完全理解。鉴于此，我们对施工图进行了详细的研究，并与同事讨论后做了一份能够清楚表达该节点的效果图——图中除了表现节点的详细结构外，还对其加以拆解和说明，方便施工人员进行施工。在其他项目中，我们也经常通过建模对设计进行反复推敲，虽然这些图并没有太多的后期，但是却对整个方案的推进起着至关重要的作用。

摄影里要有故事，效果图里更要有故事

最初，我在加入设计这行时的绘图水平并不高，也完全没有想到会在效果图上有很好的发挥。与同事互相切磋研究表现技法、参考国外不同的制作方式，甚至是危机情况的灵机一动，通过这些来提高自己的水平。而我的摄影爱好也对制作效果图起到了很大的帮助，经常会从摄影的角度去思考效果图制作——怎样才有好的画面，怎样在构图里讲故事。

李津
广州土人景观顾问有限公司 方案室 副主任设计师

It is my honor to participate in editing this book as a designer of Guangzhou Turenscape. The followings are some stories occurred in the process of making renderings during the past six years.

Crisis Moment

Various difficulties always exist in different projects, and I have never finished a project with relax. Crisis always waits for you in an unexpected corner, just like what we confronted with during a tour project in Dongguan. Before submitting the rendering, we needed to draw a moving rendering, but no one had any idea of what kind of rendering can touch the client or how to make it. At that time, Chief Pang suggested a long picture which brought striking effect like super wide screen of IMAX films. But it would take a lot of time to render such a long picture. we took SketchUp to make a quick render, together with a mount of post processing to get a fantastic effect like Avatar. Most importantly. the rendering could be finished in a short period. The client was quite satisfied with those dreamlike and impressive renderings which merged our design. From then on, we began to use SkethUp in some important projects. In this tour project, the rendering is made not only to display the design but also to create a certain atmosphere.

Various Styles

We make different styles of renderings for different project. that's why we have various renderings in this book. At first, we mainly use 3D MAX rendering, which is low efficiency since the texture and the light setting require much time to readjust. Some pictures are still more like a concept map after rendering, and need quantity of post processing to get a better effect. There is no doubt that we can get fabulous effect in this way, but it takes so much time and energy that we only adopt it in some key projects.

The SketchUp is another option, which can be finished in quite a short time. Although it presents the design contents directly and is easily recognized by people, it is our last resort for its rather crude looks. (Of course, not all pictures are so bad. It constituted a sense of graphic design in the paintings of Chengdu Qingcheng Mountain project and Guangzhou TV tower bidding project which we used SketchUp to render.) There comes the third clipping style rendering combining advantages of both above styles. It saves much time and also gets the effect of 70%~80% that 3D MAX rendering presents to make post processing on the SketchUp pictures. It does not only ensure the quality, but also increase the efficiency, and is widely used recently.

Lately, we begin to try something new that is blending the real scene with design content. It has been widely used in foreign countries long before, and we just start to try it in Guangzhou North Bank Park, Dongguan Dalingshan Project, and Dongguan Shilong Street renovation project. It is so intuitive and convenient that people can see the effect of merging design with real scence. However, it requires highly demand for making renderings that you even have to prepare the design contents when you take photos on the site, such as how to exploit the pictures taken on the scene in the rendering. The rendering should be neither too realistic nor too unrealistic, which means, those background photos to combine with design need screening and adjusting instead of using directly. The most challenging task is processing the photos to the point in terms of various demands. Bad treatment leads to a wholly disordered rendering without focus.

Actually, there are various styles of rendering, and will be more and more styles in future created not only by us but also by other colleagues.

Perform in a More Vivid Way

The last thing we want is a dead rendering of rigid practice, in which the design contents and ideal cannot be presented. We hope to shorten the distance between the design contents and viewers through rendering. How to perform vividly? We take different methods to reach this goal. When exploring the site for Dalingshan project, we took numerous photos of Location residents living, such as kids splashing in the water, old people living in peace and leisure, buffalo plowing the field. Adding those original pictures to the rendering, a vivid sight presented in front of us and the design contents seemed to already exist in real scene. Sometimes we also make little jokes by putting the hot topic figure in the rendering or sneakily putting the photo of designers into the background, which creates a humorous and funny picture.

The way of rendering composition varies in other projects, just like the super long picture we mentioned above. In certain circumstance, we make clipping or blank space for the picture, and replace the accessory objects. It creates a vivid rendering as if stepping into the picture from outside.

We need some special materials for special festival celebration scene, such as lion dance, dragon boat, spinning top and so on. These materials cannot be found in the storage, and we need to search related pictures from the internet and make some process. We prefer going out to take photos of daily life and celebrations and use the photos we take as the materials. The lifeless scene becomes lively because of those photos.

Different Functions of Rendering

Is the rendering only a blue print presented to client? The answer is obviously no. We did not make any renderings of good performance in the Shenzhen technology park project, but to our surprise, a rendering for construction guide impressed our client. The construction was undergoing at that time, but there was a certain complicated structure point which the constructor couldn't understand barely in terms of plane graph or profile map. In view of this situation, after a deep research and discussion towards the construction map, we made a rendering which not only showed the detail structure of the point but also made it clear for constructor how to manage it by illustrating in a disassembling way. In other projects, we also elaborated our design repeatedly through modeling which played a great role in boosting the whole plan without much post processing.

Epilogue

It never comes to my mind that such a poor painter like me can play well in making rendering when I first stepped into the design world. Learning from other colleagues, consulting different foreign methods, and a flash of inspiration in emergency, all these make my ability enhance. My hobby of photography is of great help for making renderings since I always stand in a photography angle to make a perfect picture—how to make a good rendering, and how to tell story in it.

Jin Li

Deputy Director of Solutions Division Designer of Guangzhou Turen Landscape CO, LTD

—

广州土人的效果图很难理出什么技法套路，倒是觉得"用不讲规矩"来形容更合适一些：如同广州土人的设计工作，没有规律可循。不同的项目，不单单是方案，连效果图的表现技法也会有所不同，这也许正是大家对设计乐此不疲的原因所在吧，而我们也认为这是对项目给出的必要的尊重。

——陈瑜璟

—

表现图的制作是设计的过程，亦是凭心中图景作画的过程，优秀的制作者是设计师也是画家。

——李娟

—

我们的图纸不华丽、不炫技，是广土众多设计师在不同时期、不同题目、不同视角下的思考。执着地用效果图直面问题、传递思考、解决问题，表明我们的设计思想和立场。

——宣传推广主任 魏敏

—

效果图是工具而非目的。设计情怀寄托于此，更寄托于设计的其它环节。精美的效果图不该为贫瘠的设计遮羞。坚持寻找一种能恰当表达设计意图与情怀的效果图表达方式，同时又执拗地试图不过分依托效果图，在当今设计行业，大概算得是"土"而"不合时宜"吧。

——副主任设计师 庹魁

—

效果图是表达设计的语言之一。但再华丽的辞藻，也不及思想。能把独到的设计思维融贯在效果图中展示并坚持着，在华而不实的效果图大行其道的今天，实属不易。

——副主任设计师 王洪诗

—

画效果图分三层：
一层——虚拟场景。表现项目竣工后的环境，材质、空间、植物、气候、光照、以及可能出现的人物，直视方案设计成果。
二层——视觉美学。角度、色彩、构图，图面本身具艺术效果。
三层——故事气氛。环境对人或事的影响，传达设计理念。

——薛轶文

—

效果图的看法？设计用词专业，我不会讲。但明显感觉其走向有点歪曲了，它应当是要真实反映出设计思想来的，光华丽漂亮怕是不行吧。我倒觉得只要双方能理解了意图理念，就是个树枝画出来的痕迹也珍贵可取。花6分力气画效果图，4分力气搞设计，这样设计岂不是就掉下去摔死。

——向瑜

广州土人景观顾问有限公司（Guangzhou Turen Landscape Planning CO.,LTD）为土人景观在华南的重要设计机构，由哈佛大学博士、北京大学教授、北京大学景观设计学研究院院长俞孔坚与北京土人景观与建筑规划设计研究院副院长庞伟创立。由庞伟先生担任总经理、首席设计师。

广州土人共享北京大学景观设计学研究院卓越的学术平台，为北京大学景观设计教学与研究实习基地。广州土人拥有独特而富有进取的工作团队，拥有各专业设计人员 50 余名，深怀知识经济时代的历史使命感，面临华南地区广大深厚的设计需求。

广州土人成立以来，多次获得国内外奖项。中山岐江公园项目荣获美国景观设计师协会（ASLA）2002 年年度最高奖项——荣誉设计奖，2003 年荣获中国建筑艺术奖，2004 年荣获第十届全国美术作品展览环境艺术类金奖及中国现代优秀民族建筑综合金奖（合作）， 2008 年荣获第 22 届世界城市滨水杰出设计"最高荣誉奖"，2009 年荣获国际城市土地学会（Urban Land Institute 简称 ULI）2009 年度 ULI 亚太区杰出荣誉大奖。广州白云国际会议中心荣获 2008 年巴塞罗那世界建筑节公共建筑设计大奖、2008 年第八届中国土木工程詹天佑大奖、2008 年广东省优秀设计奖。广东省佛山市南海区汽车城规划及建筑设计（获竞标第一名）。广东省佛山市三水西南组团中心区公共绿地景观规划设计（获优胜奖并中标）。广东省东莞市黄旗山城市公园规划设计国际竞赛（获优胜奖）。2004 年广州土人荣获广州"最佳景观设计公司"大奖。2006 年荣获"2006 年中国主流地产'金冠奖'以及最受尊敬景观建筑设计公司"称号。2007 年荣获"2007 年中国主流地产'金冠奖'以及中国地产最受尊敬景观设计公司"称号。2008 年荣获"2008 年中国主流地产'金冠奖'以及中国地产最受尊敬景观设计公司"称号。2009 年荣获"2009 年中国主流地产'金冠奖'以及年度最受尊敬景观设计公司"称号。2010 年荣获"2010 年中国主流地产'金冠奖'以及中国年度最佳景观设计公司"称号。

地址：广州番禺丽江花园德荫楼 C 座
TEL: 020-34500022 FAX: 020-84594786
EMAIL: GZTUREN@VIP.163.COM
网址：http://www.turenscape.com/gzturen/
博客：http://blog.sina.com.cn/turen09
微博：http://weibo.com/gzturen

公司荣誉

Guangzhou Turen Landscape Planning CO. LTD is founded by Kongjian Yu, doctor of Harford University, professor of Pecking University and director of the Graduate School of Landscape Architecture, Peking University, and Wei Pang, vice director of Beijing Turen Landscape and Architecture Design Institute. It has become a core design instrument of Turescape in South China, with Wei Pang as the general manager and chief designer.

Guangzhou Turenscape, a base for the teaching of landscape design and research internship for Pecking University, shares the outstanding academic platform of the Graduate School of Landscape Architecture, Pecking University. More than 50 designers from different fields compose a unique and aggressive team. They meet the broad design demand of South China with the sense of historic mission in knowledge economical time.

Guangzhou Turenscape has obtained many domestic and overseas awards since established. Zhongshan Shipyard Park won the top annual award of ASLA—Professional Award of Honor in General Design in 2002, the Chinese Architectural Art Award in 2003, Golden Metal of the tenth National Exhibition of Fine Arts and Golden Metal of excellent Modern National Architecture in 2004, Excellence on the Waterfront Top Honor Award in 2008, and ULI Award for Excellence Asia Pacific in 2009. Guangzhou Baiyun International Convention Center got the Public Architect Design Award of World Architecture Festival in Barcelona in 2008, the 8th China Civil Engineering Zhan Tianyou Award in 2008, and excellent design award of Guangdong in 2008. Motor City planning and architectural design in Nanshan District, Foshan, Guangdong province won the first place in the bidding. Landscape design of public green space for southwest group center in Sanshui Distirct, Foshan, Guangdong province won the bidding and got the winning prize. Planning for Huangqi Mt. City Park, Dongguan won the prize in international competition. Guangzhou Turenscape won the "Best Landscape Design Company" of Guangzhou in 2004. It was awarded the Gold Crown Prize of Chinese mainstream real estate and the most respected landscape and architecture design company in 2006, Gold Crown Prize of Chinese mainstream real estate and the most respected landscape design company in Chinese real estate in 2007, Gold Crown Prize of Chinese mainstream real estate and the most respected landscape design company in 2008, Gold Crown Prize of Chinese mainstream real estate and the most respected landscape design company in 2009, and Gold Crown Prize of Chinese mainstream real estate and the most respected landscape design company in China in 2010.

业务范围

Business Scope

- 开发策划

- 城市设计

- 建筑设计

- 景观设计

- 风景旅游地规划

- 城市与区域规划

- 城市与区域生态基础设施规划

- Development planning

- Urban design

- Architectural design

- Landscape design

- Scenic tourist planning

- Urban and regional planning

- Urban and regional ecological infrastructure planning

ACKNOWLEDGEMENTS / 鸣谢

感谢曾经为广州土人景观效果图表现做出贡献的同事
（以下按姓名拼音升序排列）

蔡景｜蔡智颖｜曾建洲｜陈朝东｜陈海强｜陈亮｜陈玫媛｜陈奇琦

陈思敏｜陈瑜璟｜丁涵竟｜方征｜龚文龙｜龚媛｜郭志晓｜贺建宇｜洪赟

胡电智｜胡余｜黄超力｜黄文烨｜黄征征｜季晓玲｜简言｜姜涛｜乐彭

李娟｜李理｜李帷｜李鑫｜梁瑞华｜梁诗韵｜梁天宇｜卢成军｜陆家兴

陆嘉鸿｜罗峰｜罗翔｜马轩｜蒙俊硕｜庞锋年｜裴力｜彭征｜钱克滋

邱凌凌｜庹魁｜汪乐勤｜汪耀宏｜王璋波｜魏敏｜吴懿青｜薛轶文｜杨细

杨小牧｜叶秀南｜余定｜余庆｜张健｜张康｜张燕姝｜郑驰｜朱开平｜朱涛